P9-DEX-932

ARTILLERY

GUNS & ROCKET SYSTEMS

GREENHILL MILITARY MANUALS

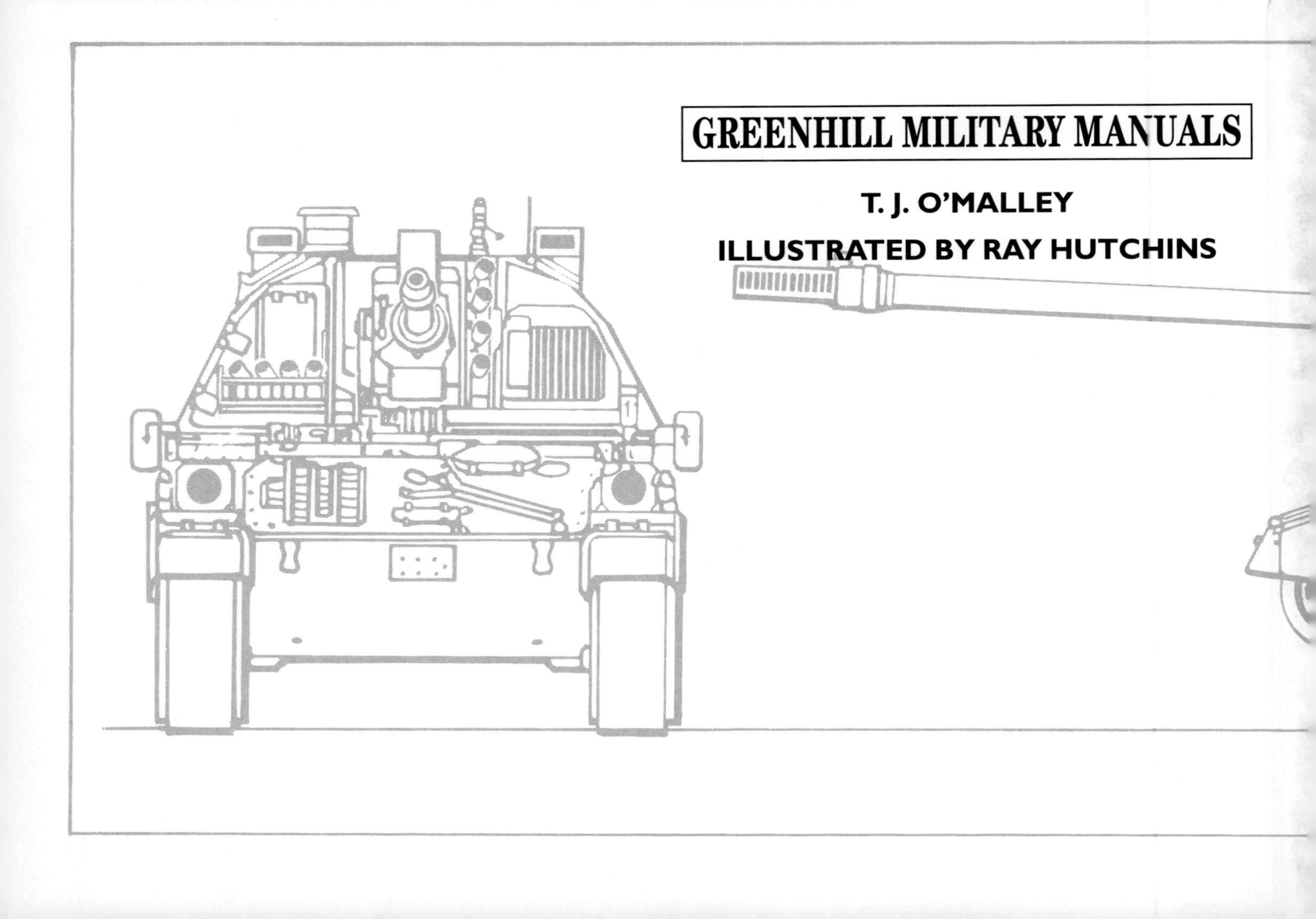

GREENHILL MILITARY MANUALS

T. J. O'MALLEY

ILLUSTRATED BY RAY HUTCHINS

ARTILLERY
GUNS & ROCKET SYSTEMS

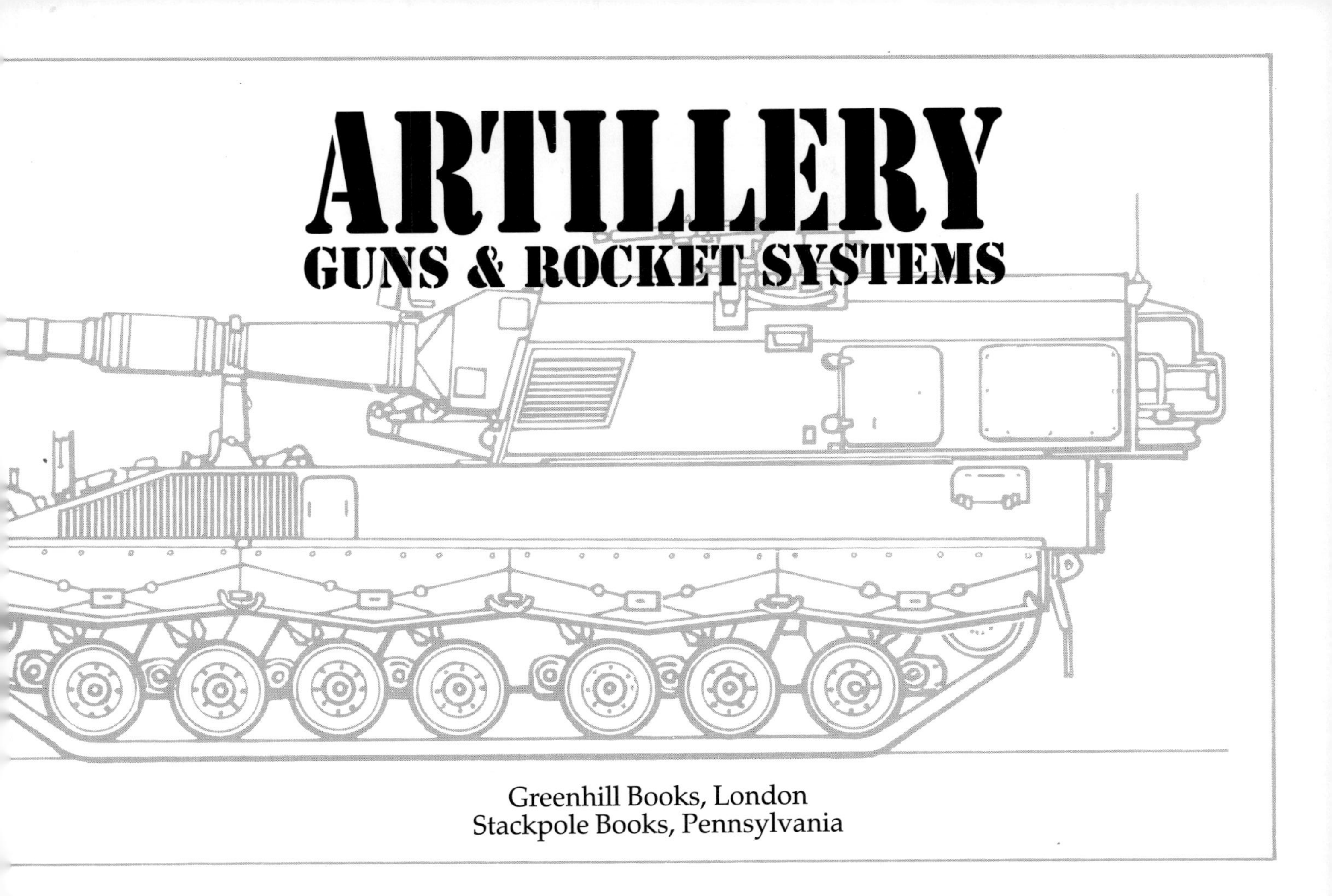

Greenhill Books, London
Stackpole Books, Pennsylvania

Artillery: Guns and Rocket Systems
first published 1994 by
Greenhill Books, Lionel Leventhal Limited, Park House
1, Russell Gardens, London NW11 9NN
and
Stackpole Books, 5067 Ritter Road, Mechanicsburg, PA 17055, USA

British Library Cataloguing in Publication Data
T.J. O'Malley
Artillery - (Greenhill Military Manuals)
I. Title II. Series 623.4
ISBN 1-85367-188-6

Library of Congress Cataloging in Publication Data
Artillery : guns and rocket systems / T.J. O'Malley
p. cm. - - (Greenhill military manuals)
ISBN 1-85367-188-6 (hc)
1. Artillery. I. Title. II. Series.
UF 145.046 1994
355.8'21 - dc20 94 - 13570
CIP

Typeset by Merlin Publications
Printed and bound in Great Britain by
Butler & Tanner Ltd. Frome and London

Introduction

No matter what new weapons and electronic-based weapon systems appear, artillery continues to dominate the battlefield. Operating as part of a team which includes infantry and armour, artillery remains the one weapon which can dictate the course of conflict, both in attack and defence.

Artillery takes many forms, but is divided mainly into towed and self-propelled types, with the quite separate categories of multiple rocket systems and heavy mortars adding to the overall potential. This survey is thus divided into those main categories but there are frequent overlaps of type and function. Of the four mentioned the self-propelled artillery piece is now the most important. Even after the addition of mobility to its firepower potential is considered the overall fact remains that the modern battlefield is a lethally hostile environment in which to survive unless armoured protection is provided. Some of the self-propelled equipments in this survey lack armoured protection for their crews, a factor soon recognised as a mistake once the 'first generation' of designs were in service. Nearly all modern self-propelled artillery designs provide armour for their crews; those that do not have crew armour rely upon their range potential providing them with the benefit of distance or the ability to 'shoot and scoot' to protect their crews from the prospect of enemy retaliation.

The towed artillery piece still has a function, usually providing fire support to formations where the weight factor overrides the provision of self-propelled artillery. This arises with airborne and similar special force formations. Large towed pieces are usually issued to reserve formations for whom the cost and maintenance loads of propelled pieces cannot be contemplated. But towed artillery systems still add their weight to the overall firepower potential.

So do the multiple rocket systems. By their very nature artillery rockets are inherently inaccurate so their multiple launch systems rely upon the delivery of large salvos to cover a target area. And cover them they do, to great effect. The huge weights of explosive that rocket systems can lay down on a target within a short period can be devastating to any target.

The heavy mortar retains its place as an artillery piece due to its relatively light weight which provides many attractions for light field and other formations. Recent developments in this area have led to an interest in self-propelled armoured mortar platforms, more of which are certain to appear in the future.

One factor that is often forgotten when any form of artillery is considered is that any artillery piece is only a delivery system in weapon system terms - the gunner's weapon is the projectile fired from the artillery piece. Hence the constant emphasis placed on ammunition in this survey.

The Future

The art of artillery seldom takes drastic strides in performance potential but the last few decades have witnessed such a stride with the innovation of the long barrel and a new generation of ammunition. Artillery barrels with lengths as long as 52 calibres are now in being, delivering operational ranges which a past generation of gunners could not have even contemplated. When these barrel length increases are coupled with range enhancement devices such as base bleed (BB) units and Enhanced Range Full Bore (ERFB) projectiles it becomes apparent that drastic changes have indeed been made. Drastic changes are still continuing. A new generation of 'smart' projectiles with self-homing warheads is with us. Coupled with the new potential ranges, a single artillery piece can now dominate huge areas of terrain on any battlefield. Gunners can now fire anti-armour projectiles in a general direction and forget about them as armoured formations way over the horizon are attacked by autonomous guided missiles at any time of the day or night. Equally large areas of terrain can be denied to an enemy by artillery delivering cargo projectiles to scatter bomblets, and land mines to prevent enemy formations manoeuvring as they might wish.

The Extended Range Full Bore (ERFB) projectile has enabled range increase margins of 30% over their conventional equivalents and these margins can be extended by the use of base bleed (BB) units in the base. This ERFB-BB example is manufactured in France by Giat Industries.

The list of innovations does not end there. Liquid propellants (LP) are almost upon us. The first system likely to utilise LP is the US Army's Advanced Field Artillery System (AFAS) which, if the funding is provided as anticipated, will start to appear in developmental form during the next few years. The logistic and safety advantages bestowed by LP will be matched by the propellant charge flexibility it will provide the gunner, doing away as it will with the existing complex systems of bagged charges and increments.

Electronics will increasingly intrude into the artillery and mortar fields, one early example being the Swedish Strix 120 mm mortar missile, seen here being programmed prior to firing; the Strix uses infra-red sensors to seek out its armoured targets.

Coupled with LP innovations there will have to be projectile handling systems. Many of these are in use already - their employment will become increasingly more widespread as personnel fatigue becomes an increasingly more significant factor in mechanised warfare.

Autonomous target-seeking sub-munitions dispensed from cargo projectiles, such as this BONUS/OBG developed by Bofors and Giat Industries, will increasingly be employed for long range attacks against armoured formations.

The anticipated advances in firepower and range potential likely to be delivered by the next generations of artillery will impose their own challenges to be overcome. Fire control will become increasingly dependent on long range sensor systems which are still in their operational infancy but no doubt electronics will find a way round.

Electronics has already demonstrated that fire control systems can place a round on target with the first shot virtually every time but those targets have to be found, assessed and engaged in a dauntingly short time. Improved fire control is the answer but it is rather outside the realm of this survey. Needless to say the gunners of the future will have to utilise systems of a complexity greater than those in service today.

Things to come? An artist's impression of the closing stages of a notional 155 mm autonomous "smart" projectile homing in against a hostile tank formation.

Contents

155 mm Type GC 45 Gun-howitzer — Belgium

Developed from 1975 onwards, the **GC 45 155 mm gun-howitzer** was the first production-standard gun designed by the late Dr Gerald Bull to fire his advanced Enhanced Range Full Bore (ERFB) streamlined projectiles and their specially developed large propellant charges. When fired from the 45-calibre barrel, itself an innovation when most contemporary pieces had 39-calibre barrels, standard ERFB projectiles could reach 30000 m. If a base bleed (BB) unit was fitted to the projectile (to make it an ERFB-BB) range could be increased to 39000 m. The first **GC 45** was fired in 1977.

The **GC 45** has a split trail carriage with a walking beam suspension for the two road wheels each side. Mounted in a carefully-balanced cradle the long slender barrel with its multi-baffle muzzle brake was manufactured using a special high-yield steel and autofrettaged throughout. The **GC 45** is not fired from its wheels but from a telescopic firing platform lowered from under the carriage.

The **GC 45s** range improvements over existing ordnance were spectacular and had a profound influence on subsequent artillery developments. However, the **GC 45**, produced by the Space Research Corporation (SRC) International of Belgium, was sold only to Thailand - the Royal Thai Marines received 12 which were produced in Canada and Austria.

With the demise of SRC International production of **GC 45** type gun-howitzers was transferred to Austria where they were produced, with some design variations, by Voest-Alpine (later NORICUM) as the **GH N-45** for exports to nations such as Thailand (a further six), Iran and Iraq. Most of the Austrian-produced examples featured an auxiliary power unit for local self-propulsion.

Although a dramatic innovation for its time the **GC 45** was really more of a development model than a fully developed production system, but it pointed the way that future artillery designs would follow, a typical example being the South African G5 (qv).

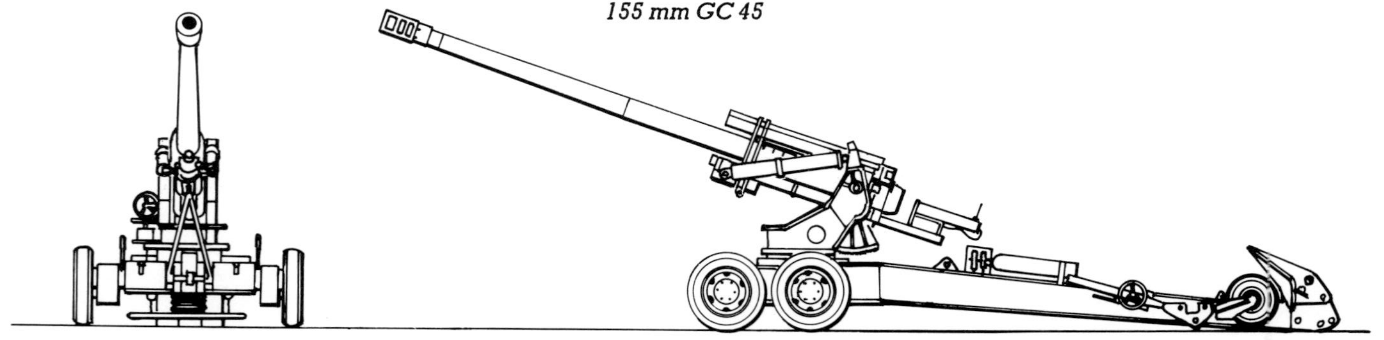

155 mm GC 45

Specification

First prototype: 1977
First production: Exact date uncertain but produced for Royal Thai Marines (12)
Current users: Thailand; Austrian-produced versions with Iran and Iraq
Crew: 8
Weight in action: 8222 kg
Barrel length: 6.975 m
Length firing: 10.82 m
Width firing: 10.364 m
Max range: ERFB 30000 m; ERFB-BB 39000 m
Muzzle velocity: 897 m/s
Projectile weight: ERFB 45.4 kg
Depression/elevation: -5°/+69°
Traverse: 80°

Austrian 155 mm GH N-45S awaiting delivery.

155 mm Type WA 021 Gun-howitzer — China

During the mid-1980s the late Dr Gerald Bull moved his design activities to the People's Republic of China where his innovative approach to artillery design was well received.

From the information imparted by Dr Bull, the China North Industries Corporation (NORINCO) produced an interim 155 mm gun-howitzer based on the Austrian GH N-45 (itself based on the SRC GC 45 - see previous entry) and known as the MH 45. From this development model NORINCO produced two further prototypes and a series of ten pre-production examples known as the **WA 021**, also referred to as the WAC 21. All these examples were produced at Heping by the end of 1986. By 1991 the **WA 021** was in service with the Chinese Army, organised into six-gun batteries.

The **WA 021** visually resembles the GC 45 and GFH N-45 but there are many detail design differences, especially in the barrel which has been produced with a number of variations in rifling twist, depth and shape, mainly to investigate the best configuration to reduce barrel wear when the larger propellant charges are fired - the production versions use rifling grooves three times deeper than other comparable barrels. Changes were introduced to the cradle to improve balance and, as with other Bull-designed 45-calibre guns, the barrel is reversed over the split trails for towing over long distances. For short distances the **WA 021** can be provided with an auxiliary power unit (APU) mounted on the forward part of the carriage. Firing a wide range of ERFB projectiles, the **WA 021** has a maximum range of 39000 metres using ERFB-BB. Projectiles available include HE, illuminating and an ERFB cargo projectile containing 72 bomblets.

The **155 mm WA 021** has been proposed by NORINCO as part of a mobile coastal defence system with up to six guns (up to 12 under emergency conditions) under the direct control of a computerised fire control station.

155 mm WA 021

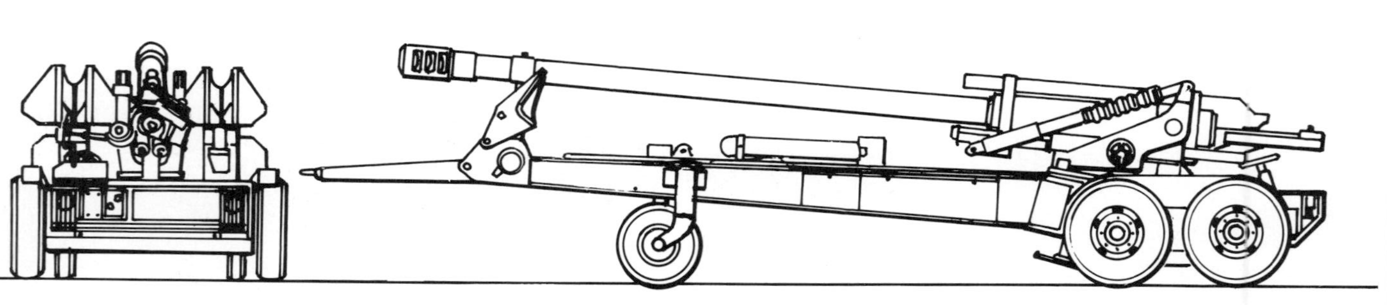

Specification

First prototype: 1986
First production: 1990-1991
Current user: People's Republic of China
Crew: 8-10
Weight in action: approx 9500 kg
Barrel length: 7.045 m
Length firing: 11.4 m
Width firing: 9.931 m
Max range: ERFB 30000 m;
ERFB-BB 39000 m
Muzzle velocity: approx 895 m/s
Projectile weight: 45.4 kg
Depression/elevation: -5°/+72°
Traverse: 30° left/40° right

155mm type WA 021 in action.

76 mm ZIS-3 Divisional Gun

The **76 mm ZIS-3** (actual calibre 76.2 mm) Divisional Gun was introduced in 1942 to supplement, and eventually replace, the large number of similar calibre artillery types then in Red Army service and to make up the huge losses inflicted by the then-advancing German armies. The **ZIS-3** was thus a simple robust design with no frills, capable of being manufactured in huge numbers running into thousands.

The **ZIS-3** is no longer an Eastern Bloc stalwart but is retained by many Third World and other armies, including China where the **ZIS-3** was locally manufactured as the 76 mm Type 54. For them the **ZIS-3** remains an important all-purpose artillery asset but it has long been withdrawn from service with the successors to the old Red Army other than as a gate guardian or saluting gun. Nations such as Romania continue to manufacture 76 mm ammunition for the **ZIS-3**, although the potential markets are dwindling.

Using a split trail carriage with tubular trail legs, a large shield and a long slender barrel with a double-baffle muzzle brake, the **ZIS-3** was intended for use as an anti-armour weapon as well as a field gun - one variant was the 57 mm ZIS-2 anti-tank gun introduced during 1943 mounted on the same carriage as the **ZIS-3**. Another variant was the self-propelled SU-76, produced in large numbers but now no longer in service anywhere.

As the **ZIS-3** was numerically more important than any other Eastern Bloc artillery piece for many years a high degree of ammunition standardisation was imposed and maintained until the present day. The most widely found is still a 6.2 kg high fragmentation HE (FRAG-HE) fixed round but other natures included various types of solid armour-piercing (AP) and, a late introduction, a fin-stabilised high explosive anti-tank (HEAT-FS-T) round capable of penetrating up to 194 mm of armour at direct fire ranges. Maximum indirect fire range was 13290 metres.

76 mm ZIS-3

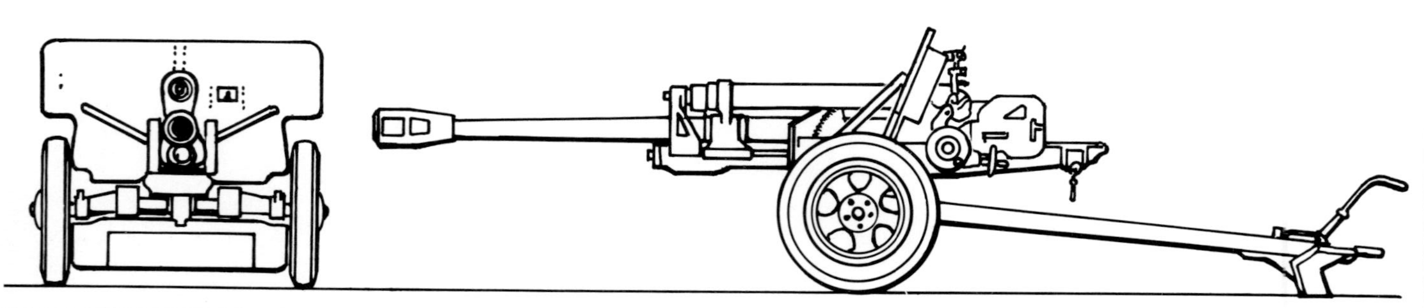

Specification
First prototype: 1942
First production: 1942
Current users: China, Romania, Cuba, Albania and many Third World nations
Crew: 6
Weight in action: 1116 kg
Barrel length: 3.455 m
Length travelling: 6.095 m
Width travelling: 1.645 m
Max range: 13290 m
Muzzle velocity: 680 m/s
Projectile weight: HE 6.2 m
Depression/elevation: -5°/+37°
Traverse: 54°

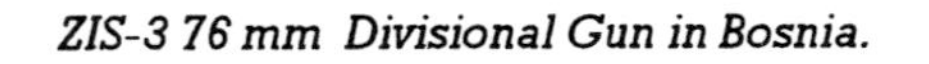

ZIS-3 76 mm Divisional Gun in Bosnia.

100 mm T-12 and MT-12 Anti-tank Guns

Former Soviet Union

Towed anti-tank guns are now comparative rarities but the old Soviet Union was a major user. Several former Warsaw Pact nations continue to employ significant numbers of a 100 mm anti-tank gun known as the **T-12**, or 2A19, and an improved version known as the **MT-12**. The **T-12** entered service during the mid-1950s and was at that time known in the West as the M1955. Service experience revealed the need for some carriage modifications and the result was the **MT-12** which appeared during 1972.

The same ordnance is used on the **T-12** and **MT-12**, a long slender 60-calibre barrel with a pepper-pot muzzle brake. The split trail carriage has a removable castor wheel close to the trail spades to assist handling. A shield is fitted. Overall the general appearance is long and low. The main difference between the two models is that the **MT-12** has a torsion bar suspension which can be locked out for firing stability.

Although the **T-12/MT-12** is intended primarily as a direct fire weapon it has indirect fire sights and can be employed as a field piece firing a HE round.

For anti-tank use both weapons can fire an armour-piercing fin-stabilised discarding sabot (APFSDS) round with a dart-like kinetic energy projectile capable of penetrating 215 mm of armour at 1000 metres. This type of round is more usually associated with tank guns but the **T-12** and **MT-12** fire a different type of fixed ammunition compared to the 100 mm D-10 series of guns used on the T-54/T-55 MBT series. Also fired by the **T-12/MT-12** is a fin-stabilised high explosive anti-tank (HEAT) round and a laser-guided projectile known as the 9M117 Kastet.

One variation of the **T-12** produced in the former Yugoslavia was created by placing of the 100 mm barrel on the carriage of the 122 mm D-30 howitzer. This variant was known as TOPAZ.

100 mm MT-12

Specification
First prototype: T-12 early 1950s; MT-12 1971
First production: T-12 1955
Current users: CIS, Iraq, Hungary, Yugoslavia
Crew: 6
Weight in action: T-12 2750 kg; MT-12 3050 kg

Barrel length: 6.126 m
Length travelling: T-12 9.5 m; MT-12 9.65 m
Width travelling: T-12 1.8 m; MT-12 2.31 m
Max range: (indirect fire) 8200 m
Muzzle velocity: APFSDS 1575 m/s; HEAT 975 m/s

Projectile weight: APFSDS 5.65 kg; HEAT 4.69 kg
Depression/elevation: -6°/+20°
Traverse: 27°

100 mm T-12 anti-tank Guns on tow by MT-LB tracked.

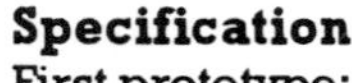

122 mm Howitzer M-30

Former Soviet Union

The **122 mm Howitzer M-30**, also known in the West as the **M1938**, is a hardy veteran, having been developed as far back as 1939 and first produced in 1939. After being produced in thousands and seeing extensive action during the Great Patriotic War the **M-30** is still in widespread service, virtually unchanged from its original form, with the CIS and many other nations, although for many armies it is now a reserve or training weapon. Although production of the **M-30** ceased in the CIS some years ago the type is still in production in China where it is known as the 122 mm Type 54 or Type 54-1, the latter having a few detail changes to suit local production methods.

The overall design of the **122 mm M-30** is completely conventional and highly robust with a split trail carriage, fixed shield with a rising centre portion, and a 23-calibre barrel without a muzzle brake - the same carriage is used by the 152 mm D-1 (M1943) howitzer (qv). The large solid wheels use sponge-filled rubber tyres although **M-30s** produced in Bulgaria have a different wheel profile. Two types of trail spade are provided with each howitzer, one for soft ground and one for hard.

The **M-30** was at one time the main armament of the SU-122 self-propelled assault gun mounted on a T-34 tank chassis but these are no longer in service anywhere. The Chinese continue to produce a self-propelled version of their Type 54-1 mounted on a Type 531 APC chassis.

The main projectile fired by the **M-30** is a very effective high fragmentation FRAG-HE weighing 21.76 kg, with a range of 11800 metres. For anti-armour operations the **M-30** can fire a BP-463 HEAT-T projectile theoretically capable of penetrating 200 mm of armour at a maximum direct fire range of 630 metres, but this projectile is now little used.

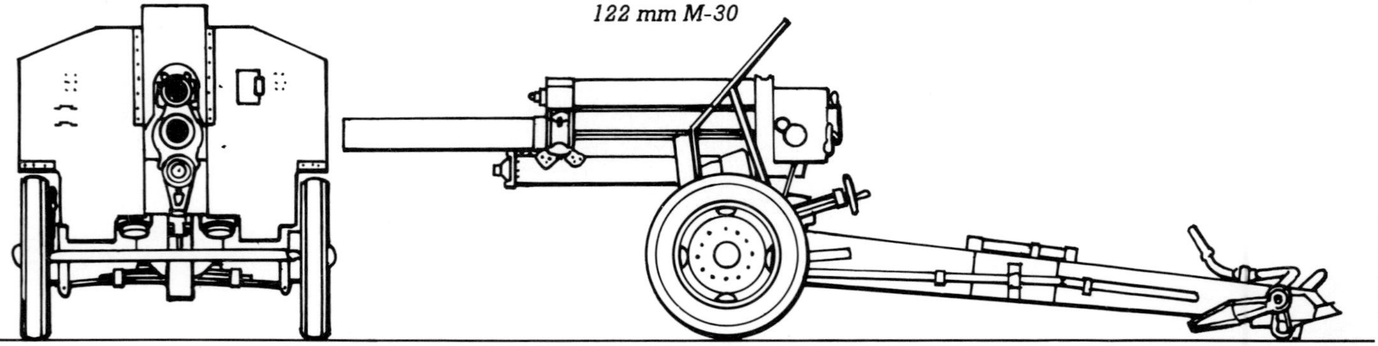

122 mm M-30

Specification

First prototype: 1938
First production: 1939
Current users: Many former Warsaw Pact nations, China and recipient nations provided with former Soviet Union military aid
Crew: 8
Weight in action: 2450 kg

Barrel length: 2.8 m
Length travelling: 5.9 m
Width travelling: 1.975 m
Max range: 11800 m
Muzzle velocity: 515 m/s
Projectile weight: 21.76 kg
Depression/elevation: -3°/+63.5°
Traverse: 49°

122 mm Howitzer M30 in firing position.

122 mm Field Gun D-74

During the late 1940s the Soviet Army had a requirement to replace their 122 mm A-19 (M1931/37) guns, a hasty expedient design produced in some numbers before 1945 but considered too heavy and short-ranged for its intended counter-battery role. In the event, the Soviet Army adopted the 130 mm Field Gun M-46 but a design produced by the Petrov design bureau for the same requirement was also manufactured as a back-up. This became the **122 mm Field Gun D-74**, first observed in 1955.

Although the Soviet Army adopted the **D-74** in limited numbers (and may still retain some for reserves), most of the production run was used for exports to spread Soviet military influence to nations such as China, Egypt, Cuba, North Korea and Vietnam. Peru also received a batch while the Chinese were sufficiently impressed to produce their own copies as the 122 mm Type 60.

The **D-74** ordnance is mounted on the same carriage as the 152 mm gun-howitzer D-20 (qv) and thus has a conventional split trail carriage with a firing platform under the forward carriage from which the gun is fired. Using this platform it is relatively quick and easy to traverse the gun through a full 360°. The long barrel, approximately 50 calibres long, is provided with a double-baffle muzzle brake. A relatively small shield is provided and castor wheels are provided on the trail legs to assist handling by the ten-man gun crew.

The **D-74** fires projectiles very similar to those used by other 122 mm systems but allied with a separate-loading variable charge cartridge case which can deliver a maximum range of 24000 metres. As usual with most CIS artillery designs, the **D-74** is intended to have an anti-armour role and thus fires a solid capped armour-piercing (APC) projectile weighing 25 kg and capable of penetrating 185 mm armour at 1000 metres. As with other similar anti-armour projectiles, this projectile is now regarded as obsolete.

122 mm D-74

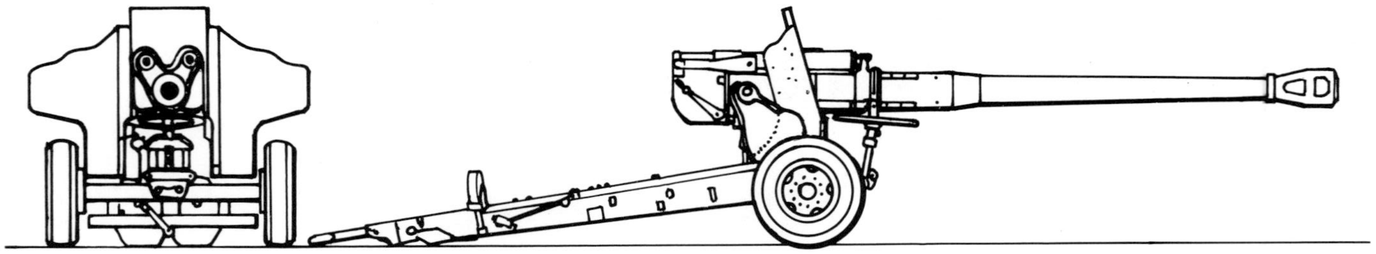

Specification
First prototype: early 1950s
First production: 1954-1955
Current users: China, Cuba, Egypt, North Korea, Peru, Vietnam
Crew: 10
Weight in action: 5500 kg
Barrel length: 6.45 m
Length travelling: 9.875 m
Width travelling: 2.35 m
Max range: 24000 m
Muzzle velocity: 885 m/s
Projectile weight: 27.3 kg
Depression/elevation: -5°/+45°
Traverse: (on carriage) 45°

122 mm Field Gun D-74 in travelling configuration.

122 mm Howitzer D-30

Former Soviet Union

The **122 mm Howitzer D-30** first appeared during the early 1960s. Its overall form, produced by the Petrov design bureau, appears to have been based on pre-1945 German designs, having a three-legged trail arrangement which allows the barrel to be rapidly traversed through a full 360° without having to move the carriage supports. The **D-30** is numerically one of the most important of all current Eastern Bloc artillery pieces and has been widely exported and licence-produced or copied in several countries. Production in the CIS has now ceased.

The **122 mm D-30** has a barrel about 38 calibres long fitted with a muzzle brake (at least two types of double-baffle brake have been produced) from which the **D-30** is towed. When emplaced the carriage wheels are raised and further protection is provided for the gun systems by a small shield.

Versions of the **D-30** are produced in China, Egypt and the former Yugoslavia (D-30J). At one time the **D-30** was manufactured in Iraq as the 'Saddam' but production there relied on imported components. Prototypes of self-propelled carriages for the Egyptian **D-30**s were produced by the American BMY and the British Royal Ordnance but the project was not continued. China produces a self-propelled version known as the 122 mm Type 85. The 122 mm ordnance used on the self-propelled 2S1 (qv) is a variant of that used on the towed **D-30**.

Late production versions of the **D-30** were known as the D-30A or 2A18M and have several detail design differences. An upgrading package is being offered by Russian establishments to improve towing speeds and time into action.

The **D-30** fires the same projectiles as the earlier 122 mm M-30 but with a larger variable propellant charge system in a cartridge case. Maximum range for FRAG-HE projectiles is 15400 metres although a rocket-assisted projectile (RAP) with a range of 21900 metres was reported at one time. Other projectiles include a high explosive anti-tank (HEAT), smoke, illuminating, leaflet and chemical (no longer used). The Chinese have developed their own family of 122 mm projectiles.

122 mm D-30

Specification

First prototype: late 1950s
First production: early 1960s
Current users: CIS, China, former Warsaw Pact nations and many others who received Soviet military aid
Crew: 7
Weight in action: 3150 kg
Barrel length: 4.875 m
Length travelling: 5.4 m
Width travelling: 1.95 m
Max range: 15400 m
Muzzle velocity: 690 m/s
Projectile weight: 21.76 kg
Depression/elevation: -7°/+70°
Traverse: 360°

120 mm Howitzer D-30 in firing position.

130 mm Field Gun M-46

Former Soviet Union

The **130 mm Field Gun M-46** was developed as a counter-battery gun to replace the old 122 mm Field Gun A-19 (M1931/37). As such it was developed during the late 1950s and was first seen in 1954. The origins of the 130 mm barrel were probably naval as the old Soviet Navy made extensive use of 130 mm guns.

The **M-46** is a bulky and heavy weapon but it has a very useful maximum range of 27150 metres, due mainly to its long barrel (length approximately 58 calibres). It is the range asset which makes the **M-46** an important weapon in many artillery parks although the type is now passing from use with many nations due to its excessive weight - when travelling the **M-46** weighs 8450 kg. Much of the weight is imparted by the bulky split-trail carriage which necessitates the use of a two-wheeled limber when on tow, usually by tracked artillery tractors - when towed the barrel is drawn back over the trails. More weight is added by the large recoil and counter-recoil mechanisms, large trail spades and a splinter-proof shield.

The Chinese value the **M-46** highly, having produced their own version as the 130 mm Type 59-1 together with an extensive ammunition family, including 130 mm ERFB-BB projectiles which increase the maximum range to 38000 metres. Nations such as Israel (an **M-46** user) and South Africa have seen fit to produce **M-46** ammunition for commercial sales. Although termed a gun the **M-46** uses a variable propellant charge system.

Despite the weight of the **M-46** only one nation developed a self-propelled carriage, namely India. The Indian Army has a number of **M-46** guns on modified Vijayanta tank chassis and known as the Catapult.

With the gradual withdrawal of the **M-46** from service it has been proposed that the carriage could accommodate a 45-calibre 155 mm barrel firing standard NATO and ERFB projectiles. Conversion packages have been offered by Israel, China and, at one time, the former Yugoslavia.

130 mm M-46

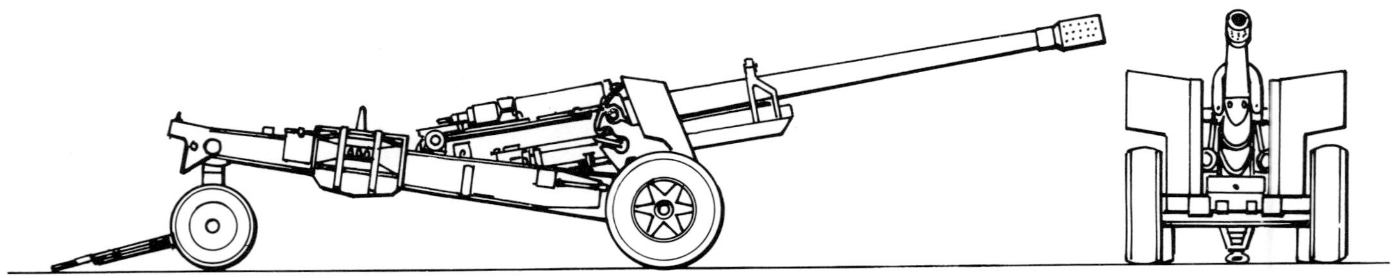

Specification
First prototype: early 1950s
First production: 1954(?)
Current users: some 40 countries, including the CIS, China, Egypt and India
Crew: 8-10
Weight in action: 7700 kg
Barrel length: 7.6 m
Length travelling: 11.73 m
Width travelling: 2.45 m
Max range: 27150 m
Muzzle velocity: 930 m/s
Projectile weight: (FRAG-HE) 33.4 kg
Depression/elevation: -2.5°/+45°
Traverse: 50°

130 mm Field Gun M-46 on parade in Egypt.

152 mm Howitzer D-1

The **152 mm Howitzer D-1** (actual calibre 152.4 mm) is often referred to as the M1943 from the year of its introduction into service with the old Red Army. Developed as a wartime expedient, again by the highly active Petrov design bureau, the **D-1** continued an established Soviet practice of combining components from two existing artillery pieces. The split trail carriage, shield and recoil system of the 122 mm Howitzer M-30 (qv) were combined with the ordnance of the152 mm Howitzer M-10 (M1938), along with the M-10 variable charge ammunition family. Where necessary the M-30 carriage was strengthened to accommodate the extra stresses and weights involved.

The result was a highly successful howitzer nearly 1000 kg lighter than the over-heavy 152 mm Howitzer M-10 which the **D-1** replaced in production at several locations. The weight decrease made the **D-1** much easier to handle yet it retained the same maximum range (12400 metres) and on-target power as the heavier design. In fact the **D-1** was so successful that it is still retained in service, albeit mainly for training and reserves, with the CIS and several other ex-Warsaw Pact states and with nations such as China (who had their own virtually identical version, the 152 mm Type 54), Iraq, Cuba and Egypt (among others).

The main projectile fired by the **D-1** is a very effective FRAG-HE weighing 40 kg and containing nearly 6 kg of TNT, although many other types of projectile have been developed over the years, including screening smcke and an associated chemical projectile (filled with Lewisite) which has now been withdrawn. Other projectiles have included a special concrete-piercing shell for use against fortifications or urban targets and an anti-armour high explosive anti-tank (HEAT), also now withdrawn other than by Romania where a locally-designed fin-stabilised HEAT round is still being produced.

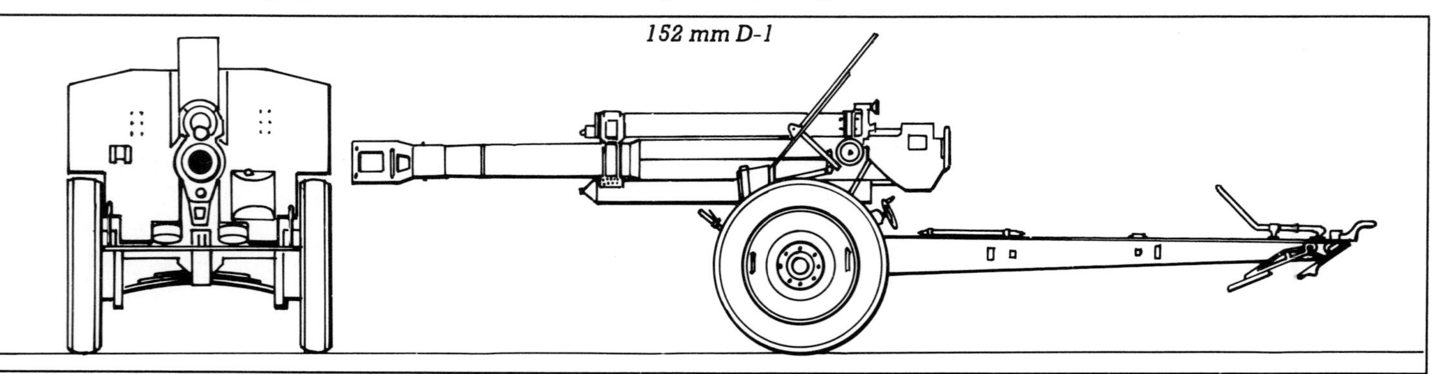

152 mm D-1

Specification

First prototype: 1943
First production: 1943
Current users: CIS, China, Afghanistan, Iraq, Hungary, Mozambique, Syria, Vietnam, Cuba, Albania and others
Crew: 7
Weight in action: 3600kg
Length travelling: 7.558 m
Barrel length: 4.207 m
Length travelling: 7.558 m
Width travelling: 1.994 m
Max range: 12400 m
Muzzle velocity: 508 m/s
Projectile weight: 40 kg
Depression/elevation: -3°/+63.5°
Traverse: 35°

152 mm D-1s in action.

152 mm Gun-howitzer D-20

The **152 mm Gun-howitzer D-20** is another example of the old Soviet practice of combining components from existing artillery systems into a new form, in this case a new 152 mm ordnance with the 122 mm Field Gun D-74 (qv) carriage. The 152 mm gun-howitzer was developed immediately after the Great Patriotic War (1941-1945) but was not placed into production until during the early 1950s to produce the **D-20**, first seen in 1955.

The **D-20** is basically similar to the earlier 152 mm Howitzer D-1 but the arrangement of the recoil cylinders on the **D-20** is different while the bulk of the split trail D-74 carriage makes the **D-20** a heavier weapon to the extent that castor wheels under the trail legs are necessary. Other differences can be seen in the shape of the shields.

Where the **D-20** differs mainly from the D-1 is in the scope of the ammunition available. The **D-20** can still fire most D-1 projectiles but the **D-20** uses a new family. The way ahead was signified by the **D-20** being one of the first Soviet artillery pieces to be capable of firing a tactical nuclear projectile. Other **D-20** ammunition natures included various chemical-filled projectiles (now withdrawn). A revised variable charge system was introduced to increase the maximum range of the **D-20** to 17410 metres while a rocket-assisted projectile (RAP) increases the range potential to 24000 metres. A more recent innovation is the laser-guided Krasnopol anti-armour projectile which weighs 50 kg when fired.

The 152 mm ordnance used on the 2S5 self-propelled howitzer is a variant of that used with the **D-20**. The former Yugoslavia was offering a 39-calibre version of the **D-20** which was apparently accepted by the former Yugoslav Army - its present status is uncertain. Romania has in service a locally-developed towed howitzer known as the Model M1985 which demonstrates some **D-20** features. The Chinese-manufactured version is the 152 mm Type 66.

152 mm D-20

Specification
First prototype: late 1940s
First production: 1954 or 1955
Current users: CIS, China, Afghanistan, Algeria, Ethiopia, Egypt, Hungary, Nicaragua, India and others
Crew: 10
Weight in action: 5650 kg

Barrel length: 5.195 m
Length travelling: 8.69 m
Width travelling: 2.32 m
Max range: 17410 m; RAP 24000 m
Muzzle velocity: 655 m/s
Projectile weight: FRAG-HE 43.51 kg
Depression/elevation: -5°/+63°
Traverse: 58°

152 mm Type 66, the Chinese version of the 152 mm Gun-howitzer D-20.

152 mm Howitzer 2A65

The **152 mm Howitzer 2A65** may be regarded as the latest of the long line of 152 mm field howitzers of Soviet and now CIS origin. As yet relatively few details regarding the **2A65**, also known in the West as the M1987 (another designation used is MSTA-B), have been released even though the type has been actively marketed for possible export sales.

There are few dramatic innovations on the **2A65**, the most obvious changes compared to earlier howitzers being the long slender barrel which is estimated to be about 40 calibres long - a double-baffle muzzle brake is fitted. The split trail carriage has few features of note other than swivelling castor wheels secured towards the end of each trail leg and the provision of a hydraulically raised and lowered firing platform under the forward carriage. Getting the **2A65** in and out of action takes the eight-man crew from 2 to 2.5 minutes. A ramming mechanism and a semi-automatic breech can produce a maximum rate of fire of seven rounds a minute. Towing speeds on roads can be as high as 80 km/h or 20 km/h across country.

The **2A65** can fire most existing 152 mm howitzer projectiles but a new family has been developed specifically for the **2A65** and similar howitzers. In this family one of three pre-loaded cartridge cases can be selected to suit a particular fire mission. Firing the latest FRAG-HE projectile the maximum range is 24000 metres but a FRAG-HE with a base bleed (BB) unit added can improve this to 29000 metres. Other 152 mm ammunition innovations introduced with the **2A65** include a cargo round containing 42 dual purpose (anti-personnel and armour-penetrating) bomblets, each containing 45 grams of explosive. Another novel projectile dispenses small radio transmitter bodies intended to jam enemy communications. The **2A65** can also fire the laser-guided Krasnopol anti-armour missile.

The ordnance of the **2A65** is essentially similar to that used on the self-propelled 2S19 gun (qv).

152 mm 2A65

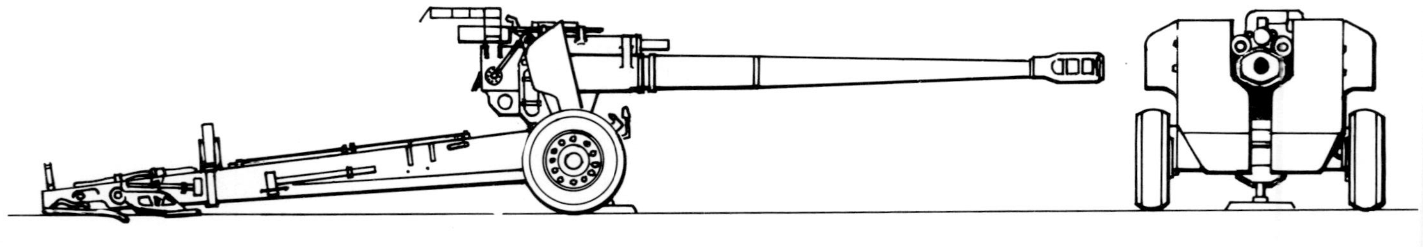

Specification
First prototype: early 1980s
First production: 1986(?)
Current user: CIS
Crew: 8
Weight in action: 7000 kg
Max range: FRAG-HE 24000 m;
FRAG-HE-BB 29000 m
Muzzle velocity: 828 m/s
Projectile weight: 42.86 kg
Depression/elevation: -3.5°/+70°
Traverse: 52°

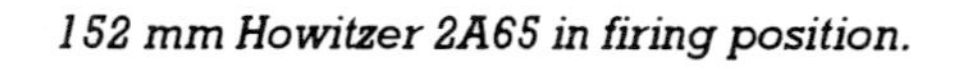

152 mm Howitzer 2A65 in firing position.

152 mm Gun 2A36

Former Soviet Union

With the gradual withdrawal of the 130 mm Field Gun M-46 (qv) the old Soviet Army introduced a requirement for a new long-range counter-battery gun. Two 152 mm systems were developed, one being the self-propelled 2S5 (qv) and the other the towed **2A36**, both using virtually identical ordnance. The towed **2A36** was originally known in the West as the 152 mm M1976 from the year it was first observed - the CIS armed forces use the code name Giatsint (Hyacinth).

Unlike many earlier Soviet artillery developments the **2A36** is entirely new. The barrel is 49 calibres long with a multi-baffle muzzle brake while the split trail carriage travels on a walking beam suspension involving two road wheels each side. In the firing position the carriage of the **2A36** rests on a circular platform lowered from under the forward carriage. One unusual feature on Soviet artillery is a hydraulically-powered load-assist device combined with an automatic horizontally-sliding breech block. This combination enables the **2A36** to fire up to six rounds a minute so that a single eight-gun **2A36** battery can place almost one tonne of projectiles onto a target in that one minute. As is usual on Soviet artillery designs the **2A36** has a shield to protect the gun systems while protection is also provided for the recoil cylinders over the barrel.

The long barrel of the **2A36** produces a high muzzle velocity of approximately 800 m/s and a correspondingly long range - 27000 metres firing standard projectiles. A reported 40000 metres is achievable using a rocket-assisted projectile (RAP). The **2A36** fires a family of streamlined separate-loading ammunition shared only by the self-propelled 2S5 including, at one period, a tactical nuclear projectile. Other rounds include a FRAG-HE (the most widely used, weighing 46 kg), concrete-piercing, incendiary, and chemical (now withdrawn). For direct fire use against armoured targets the **2A36** fires an armour-piercing round.

152 mm 2A36

Specification

First prototype: early 1970s
First production: 1975-1976
Current users: CIS, Finland and Iraq
Crew: 8
Weight in action: 9760 kg
Barrel length: 8.197 m
Length firing: 12.3 m
Width travelling: 2.788 m
Max range: FRAG-HE 27000 m; RAP 40000 m
Muzzle velocity: approx 800 m/s
Projectile weight: FRAG-HE 46 kg
Depression/elevation: -2.5°/+57°
Traverse: 50°

152 mm Gun 2A36 on display in St Petersburg.

180 mm Gun S-23

Despite having been first observed during 1955 the origins of the **180 mm Gun S-23** are still obscure. The most probable origin of the **S-23** was a naval or coast defence gun converted for the long range heavy artillery role. For many years so little was known regarding the **S-23** that it was long known in Western artillery circles as the 203 mm M1955. It was not until examples captured during one of the Middle East conflicts were examined that the actual calibre was discovered to be 180 mm.

The **S-23** is a bulky and heavy weapon weighing nearly 21.5 tonnes when emplaced. The shape of the 48-calibre barrel and the location of its associated recoil mechanisms would seem to indicate naval origins and the large breech block has a screw mechanism, although the pepper-pot muzzle brake is certainly non-naval. There is no shield and the barrel is mounted on a heavy split trail carriage which has to be carried on a wheeled limber when on tow-the usual towing vehicle is a heavy tracked tractor. For firing the **S-23** rests on a firing platform lowered from under the front carriage. When on the move the **S-23** barrel travels on dual solid-tyred wheels and the barrel is drawn back to rest over the trail legs.

The usual projectile fired by the **S-23** is a FRAG-HE weighing 88 kg, of which over 10 kg is explosive. The maximum range of this round is 30400 metres but a RAP version with a smaller explosive payload can reach 43800 metres. Other projectile available include a concrete-piercing shell intended for the demolition of fortifications and other heavy structures. Variable bagged propellant charges are used with the **S-23**.

The **180 mm S-23** was apparently never produced in large numbers and the type has been largely withdrawn from CIS service. Some were exported to nations such as India and Syria but their current status with nations such as Iraq is now uncertain.

180 mm S-23

Specification

First prototype: early 1950s(?)
First production: 1953(?)
Current users: CIS, India, Egypt, Iraq (?), Somalia and Syria
Crew: 16
Weight in action: 21450 kg
Barrel length: 8.8 m
Length travelling: 10.485 m
Width travelling: 2.996 m
Max range: FRAG-HE 30400 m; RAP 43800 m
Muzzle velocity: FRAG-HE 790 m/s
Projectile weight: FRAG-HE 88 kg
Depression/elevation: -2°/+50°
Traverse: 44°

180 mm Gun S-23 in travelling configuration.

Giat 105 mm LG1 Light Gun

France

The **105 mm LG1 Light Gun** was developed as a private venture by Giat Industries to provide potential markets with an artillery piece suitable for use by special forces operating in difficult terrain. Three prototypes were produced by 1987, with production following an order placed by Singapore in 1990 for 'at least' 36 examples, the last of which was delivered in 1991. Singapore operates the **LG1** in two battalions, each with three six-gun batteries.

The **LG1** has a 30-calibre barrel with a double-baffle muzzle brake and uses a split trail carriage. One unusual feature is the small shield fixed to the ordnance - the shield is raised and lowered with the barrel. To facilitate opening and closing the trails a hand-operated hydraulic pump is provided. Firing is from a circular firing platform located under the front carriage. Getting the **LG1** in and out of action is stated to take only 30 seconds. The usual tractor vehicle is a light 4 x 4 truck and for towing over long distances the barrel can be reversed over the trails to make the towed load more stable when crossing uneven terrain.

The **LG1** can fire NATO standard 105 mm ammunition - the HE M1 projectile can be fired to a range of 11500 metres. To make full use of the 30-calibre barrel Giat developed special base bleed (BB) projectiles with a maximum range of 17500 metres. Ammunition types include HE and smoke in both base bleed and boat-tailed (BT) versions, the latter having a maximum range of 15000 metres. It is possible to fire these rounds, which have a higher payload capacity than the NATO projectiles, from other long-barrelled 105 mm howitzers, including an updated version of the American 105 mm M101A1 (qv) using the same barrel as the **LG1**. This conversion is also produced by Giat Industries and has been produced for an 'undisclosed customer', believed to be Thailand.

105 mm LG1

Specification

First prototype: 1986-1987
First production: 1990
Current user: Republic of Singapore
Crew: 6
Weight in action: 1485 kg
Barrel length: 3.15 m
Length firing: 6.76 m
Width firing: approx 3.5 m
Max range: (Giat BB) 17500 m
Muzzle velocity: n/av
Projectile weight: (Giat HE BB) 13 kg
Depression/elevation: -3°/+70°
Traverse: 36°

Giat 105 mm LG1 Light Gun on training exercise.

Giat 155 mm Towed Gun TR

France

The **Giat 155 mm Towed Gun TR** was developed specifically to provide the French Army's motorised infantry divisions with a modern towed artillery piece to replace all existing artillery in service. The first prototype was demonstrated in 1979 but development progress was slow so it was not until late 1987 that troop trials commenced. Production commenced in 1989 with the current total expected to be around 100.

The **155 mm TR** has a 39-calibre barrel which is ballistically matched for firing NATO ammunition although the French Army make use of their their own Giat-developed propelling charge system and projectiles. Standard NATO rounds can, however, be fired from the **TR**. Giat 155 mm projectiles include cargo projectiles carrying either 63 bomblets or six anti-tank mines.

The **TR** barrel is mounted on a split trail carriage with carriage services, such as opening and closing the trail legs, provided by a hydraulic system powered from an auxiliary power unit (APU) mounted on the front carriage. The APU also provides power for short moves, although for long moves a 6 x 6 truck is used, with the **TR** barrel reversed over the trail. Hydraulic power ramming is employed to produce consistent firing results.

Maximum range firing a special HE base bleed (HE BB) projectile is 32000 metres although an increase to 41500 metres can be achieved (using ERFB-BB) if a proposed 52-calibre barrel is introduced. A **TR** prototype with a 52-calibre barrel was demonstrated in 1990 but its future is uncertain. Another **TR** version with a 45-calibre barrel was produced, also in 1990, having a maximum range of 39500 metres, again using ERFB-BB.

The barrel of the **TR** has been proposed as a replacement for the barrel of the American 155 mm M114 towed howitzer. Known as the 155 mm M114F, the modified M114 would have a similar ballistic performance to the standard **TR**, and features several other updating measures other than the longer **TR** barrel.

155 mm Giat TR

Specification
First prototype: 1979
First production: 1989
Current user: French Army
Crew: 7
Weight in action: approx 10750 kg
Barrel length: 6.2 m
Length firing: 10 m
Width firing: 8.4 m
Max range: (HE BB) 32000 m
Muzzle velocity: 830 m/s
Projectile weight: 43.5 kg
Depression/elevation: -6°/+66°
Traverse: 65°

Giat 155 mm Towed Gun TR in firing position.

105 mm Model 56 Pack Howitzer

Italy

The OTO Melara **Model 56 pack howitzer** has been one of the most successful of post-war Italian artillery designs, having been in production since 1957 and exported to more than 30 countries, including the United Kingdom, where the **Type 56** formed the main strength of the Royal Artillery's towed batteries for many years (they are still used for training). Production is now on an 'as required' basis at OTO Melara's facility at La Spezia.

The **105 mm Type 56** has several unusual features not the least of which is a variable length split trail carriage which can be used in a standard two-section form for normal indirect firing or in a lowered three-section high stability form for anti-tank operations - the raising and lowering is achieved using angled stub axles. For towing, the trail legs are folded up and over, with jacks keeping the carriage stable until the towing vehicle, usually a light 4 x 4 Jeep-type vehicle, is connected. If required the **Type 56** can be towed by one or two draught animals, including oxen and camels.

The **Type 56** can be broken down into 11 main sub-assemblies for pack transport either by animal, helicopter, vehicle or troops - the heaviest load weighs 122 kg. Stripping the **Type 56** for pack transport takes a trained crew three minutes, with reassembly taking four minutes. If required, the two-part shield can be left off to reduce weight.

The **Type 56** fires standard NATO 105 mm HE M1 semi-fixed rounds to a maximum range of 10575 metres from its 14-calibre barrel. For anti-tank operations the HEAT M67 projectile is fired to penetrate 102 mm of armour at direct fire ranges. To reduce recoil forces the standard **Type 56** is provided with a multi-baffle muzzle brake, although examples used by the German Army feature a large single-baffle muzzle attachment.

105 mm Model 56

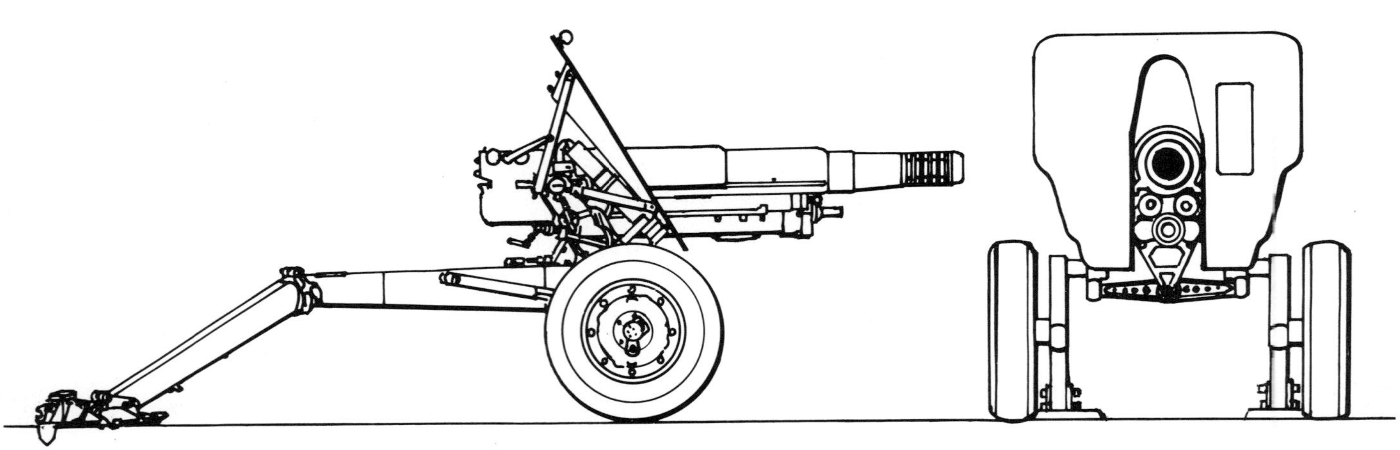

Specification

First prototype: 1956
First production: 1957
Current users: over 30 countries
Crew: 6 or 7
Weight in action: 1290 kg
Barrel length: 1.478 m
Length firing: 4.8 m
Width firing: 2.9 m
Max range: 10575 m
Muzzle velocity: 472 m/s
Projectile weight: 14.97 kg
Depression/elevation: -5°/+65°
Traverse: 36°

105 mm Model 56 Pack Howitzer at maximum elevation.

105 mm KH 178 Light Howitzer

South Korea

Following the examination of examples of the British Royal Ordnance 105 mm Light Gun (qv) and a German Rheinmetall-produced update of the American 105 mm M101 howitzer, the Kia Machine Tool Company undertook their own in-house development of a modernised version of the venerable M101. Production of the new model, the **105 mm KH178**, for the South Korean Army commenced in 1984. Since then attempts have been made to export the **KH178**, without apparent success.

The **KH178** may be regarded as a thoroughly updated version of the standard M101 howitzer fitted with a 34-calibre barrel with a double-baffle muzzle brake. Also revised are the recoil system, the progressive twist rifling, the fire control system and the horizontal sliding block breech mechanism. Alterations were also introduced to the barrel elevating and balancing mechanism to compensate for the increased weights and recoil forces, to the extent of allowing a maximum rate of fire of up to 15 rounds a minute. A splinter-proof shield is optional although it seems to be fitted to most in-service examples. Many components of the original M101 howitzer continue to be used unchanged although many items, such as the split trail legs, are strengthened. Run-flat tyres are fitted as standard.

The **KH178** continues to fire standard 105 mm ammunition of the HE M1 semi-fixed type although the longer barrel permits the use of the M200 single-increment supercharge cartridge which produces a maximum range of 14700 metres. Using rocket assisted projectiles (RAP, also known as HERA - high explosive rocket assisted), such as the M548, the maximum range is increased to 18000 m. Also fired are M327 HEP projectiles for anti-armour operations. The **KH178** is one of the few artillery pieces still intended for use with the M546 anti-personnel projectile which breaks open to release 8000 tiny flechettes with devastating results at both long and short ranges.

105 mm KH 178

A rare photograph of the 105 mm KH 178 Light Howitzer.

Specification
First prototype: early 1980s
First production: 1984
Current user: South Korean Army
Crew: 6 to 8
Weight in action: 2650 kg
Barrel length: 4.48 m
Length firing: 7.56 m
Width: (towing) 2.1 m
Max range: HE 14700 m; HERA 18000 m
Muzzle velocity: HE 662 m/s
Projectile weight: 14.97 kg
Depression/elevation: -5°/+65°
Traverse: 45.5°

155 mm KH179 Howitzer

South Korea

The South Korea Kia Machine Tool Company were involved in one of the first successful attempts to update the venerable American 155 mm M114 towed howitzer, with their development programme commencing in 1979. A prototype was ready by 1982 and production of the **KH179** for the South Korean Army began in 1984. Production is now on an 'as required' basis. It should be noted that although the **KH179** is an update of the M114, most of the **KH179** is produced from new.

The main change introduced on the **KH179** is a new 39-calibre barrel fitted with a prominent double-baffle muzzle brake. The new monobloc barrel is made from special high strength alloy steel and is used together with a revised interrupted screw breech mechanism which limits the maximum rate of fire to about four rounds a minute. Most of the other changes introduced on the **KH179** are concerned with the extra weight and recoil forces introduced by the new barrel and include recoil and balancing system changes plus modifications to the gun controls. New sights are provided.

The **KH179** is light enough to be carried slung under a CH-47D helicopter although the more usual mode of movement is behind a heavy 6 x 6 truck at speeds of up to 70 km/h. The firing position involves a plate located under the main axle, with the plate lowered for firing so that the **KH179** is stabilised on the plate and trail spades only.

The new barrel of the **KH179** continues to accept all NATO standard 155 mm ammunition, including the various ICM (Improved Conventional Munition) cargo rounds and anti-armour projectiles, and including the laser-guided Copperhead, understood to be in service with the South Korean Army in undisclosed numbers. It is also possible to fire ERFB projectiles, although South Korean batteries usually make use of the M549 HERA (high explosive rocket assisted) projectile which can reach a range of 30000 metres. The standard range firing 155 mm HE M107 is given as 22000 metres.

155 mm KH 179

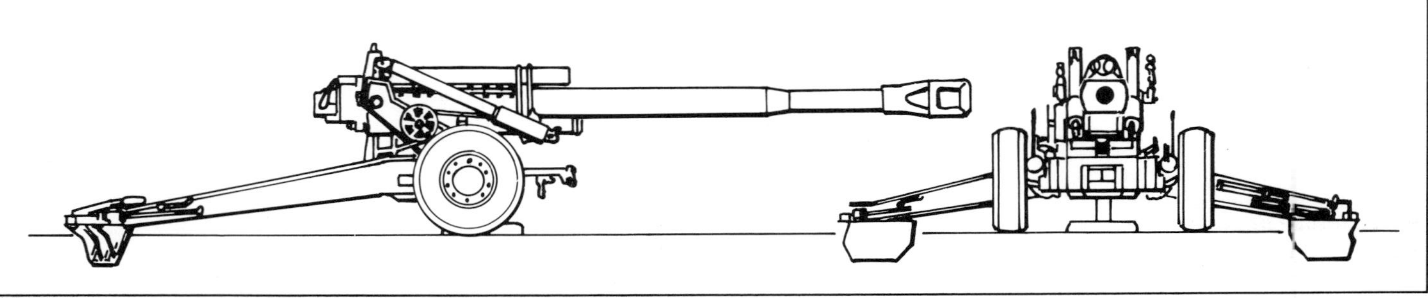

Specification
First prototype: 1982
First production: 1984
Current user: South Korean Army
Crew: 10 or 11
Weight in action: 6890 kg
Barrel length: (total) 7.013 m
Length firing: 9.55 m
Width firing: 5.05 m
Max range: HE 22000 m; HERA 30000 m
Muzzle velocity: 826 m/s
Projectile weight: HE 43.9 kg
Depression/elevation: 0°/+68.6°
Traverse: 48.7°

A clandestine photograph of the 105 mm KH 179 Howitzer in firing position.

RDM 155 mm M139/39 & M139 Howitzers

Netherlands

The Dutch RDM BV of Rotterdam were another engineering concern to realise the conversion potential of the venerable American 155 mm M114 towed howitzer. By replacing the old 23-calibre barrel with a new 39-calibre barrel the converted howitzers would be able to fire modern ammunition to greatly increased ranges at a much lower cost than procuring comparable performance artillery systems from new.

RDM named their conversion the **155 mm M139/39**. From the outset RDM decided to retain as many parts of the existing M114s as possible but allied them with a new 39-calibre barrel manufactured for RDM by the Swedish Bofors AB. The intention was originally to produce a barrel ballistically similar to that used on the 155 mm M109A2 series of self-propelled howitzers but the changes also meant that the new barrels could fire ERFB and ERFB-BB projectiles with their enhanced ranges. The introduction of the longer barrel with its multi-baffle muzzle brake necessitated some modifications to the M114 recoil and balancing mechanisms, plus the relocation of the firing plate to a position further forward on the carriage. Many more detail changes were also introduced, including modifying the trails to assist handling. Provisions for optional extras such as a power rammer were also made.

All these modifications can either be carried out by RDM or issued to a user in kit form. So far the Netherlands, Canada, Denmark and Norway have ordered the **M139/39** option. RDM can also produce completely all-new howitzers, virtually identical to the converted **M139/39**, known as the **M139**.

Both the **M139/39** and **M139** can fire all NATO standard 155 mm ammunition, with the HE M107 reaching 18200 metres. If ERFB-BB projectiles become involved the maximum range is improved to 32400 metres but, no doubt for reasons of economy, many users continue to employ the NATO standard projectiles and propellants.

155 mm M139/39

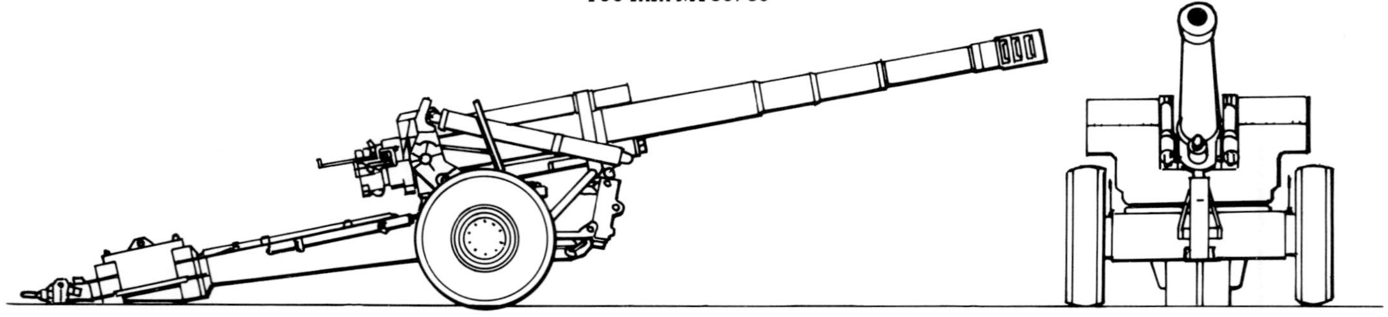

Specification

First prototype: 1984
First production: 1987
Current users: Netherlands, Denmark, Norway. Ordered by Canada
Crew: 11
Weight in action: 7600 kg
Barrel length: 6.016 m
Length: (travelling) 10 m
Width: (travelling) 2.44 m
Max range: HE M107 18200 m; ERFB-BB 32400 m
Muzzle velocity: HE M107 684 m/s
Projectile weight: HE M107 43.9 kg
Depression/elevation: -2°/+63°
Traverse: 49°

RDM 155 mm M139/39 Howitzer on tow.

ODE 155 mm FH-88 Gun-howitzer

Singapore

The Republic of Singapore is not a nation normally associated with artillery development but the government-owned Ordnance Development and Engineering of Singapore (ODE) decided during the early 1980s to develop a modern 155 mm towed gun-howitzer. Using design analysis gleaned from many sources the result was a prototype produced in 1983. Design modifications were introduced as a result of trials with a series of prototype and pre-production standard gun-howitzers before production of the finalised design, the **FH-88**, commenced during 1987. The following year the **FH-88** entered service with the Singapore Army.

The **155 mm FH-88** has a 39-calibre barrel mounted on a split trail carriage powered by a 96 hp auxiliary power unit (APU). While closely following many established design trends introduced elsewhere the **FH-88** has many design niceties of its own, including many features intended to enable the **FH-88** to be handled easily by a crew numbering only six. To this end the APU powers hydraulic circuits for opening and closing the trail legs and raising and lowering the trail dolly wheels. Batteries and hand pumps are provided as back-ups in the case of an APU failure. The APU can also power the **FH-88** over short distances - most moves are made with the barrel reversed over the trail legs.

A flick rammer is provided for a maximum fire rate of three rounds in 15 seconds. Many of the components used to manufacture the **FH-88** are imported as forgings and machine finished in Singapore. High strength alloy steels are used throughout.

The **FH-88** can fire all standard NATO 155 mm ammunition - the HE M107 can reach 19000 metres while a hollow base ERFB-HB projectile (also produced in Singapore) can reach 24000 metres.

In 1990 ODE introduced a prototype 52-calibre version of the **FH-88** as a private venture. Using ERFB-BB the 52-calibre **FH-88** could have a maximum range potential of over 40000 metres.

155 mm FH-88

Specification

First prototype: 1983
First production: 1987
Current user: Singapore
Crew: 6
Weight in action: 12800 kg
Barrel length: 6.1 m
Length firing: 9.88 m
Width firing: 8.2 m
Max range: HE M107 19000 m;
ERFB-HB 24000 m
Muzzle velocity: approx 825 m/s
Projectile weight: HE M107 43.9 kg
Depression/elevation: -3°/+70°
Traverse: 60°

ODE 155 mm FH-88 Gun-howitzer in action.

155 mm G5 Towed Gun-howitzer

South Africa

As a result of the 1970s United Nations arms embargo on South Africa, a concern known as Armscor was established to supervise defence development and procurement in South Africa. As their in-service artillery was at that time out-ranged by the ex-Soviet artillery used by many of the nations bordering on South Africa, artillery development was high on the Armscor agenda and the 155 mm GC 45 (qv) developed by SRC International in particular.

Armscor obtained the means to develop the GC 45. By a steady process of trials and their own innovative skills, Armscor engineers were able to develop the original Bull-inspired design to a point where it entered service as the **155 mm G5** in 1983. Since then the **G5** has been further developed as one component in an artillery system now generally regarded as one of the best in the world. Exports have been made to Iraq and Qatar.

The **G5** now hardly resembles the GC 45 original as it was developed to a point where it is a much more serviceable howitzer. Changes introduced in South Africa include a much heavier 45-calibre barrel, a better-balanced cradle and carriage, strengthened components throughout, a new muzzle brake, and a 79 hp auxiliary power unit (APU) to power the trails and trail wheels and provide power for self-propulsion.

The **G5** is part of an artillery system which includes fire control, meteorology stations, crew communications and an ERFB-based ammunition family. ERFB projectiles can be converted to ERFB-BB and back again by the addition or removal of a base bleed unit in the field. Ammunition types include HE, illuminating, cargo, radar chaff, smoke and even leaflet dispensers. The **G5** ordnance and ammunition system is compatible with that used on the self-propelled 155 mm G6 (qv). Maximum ranges for ERFB-BB projectiles are over 40000 metres under some South African climatic conditions.

A 52-calibre barrel for the **G5** was introduced in prototype form during 1992. This has a potential ERFB-BB range of over 42200 metres.

155 mm G5

155 mm G5 Towed Gun-howitzer firing at low elevation angle.

Specification
First prototype: late 1970s
First production: 1982
Current users: South Africa, Iraq, Qatar
Crew: 5 or 6
Weight in action: 13750 kg
Barrel length: 6.975 m
Length firing: 11 m
Width firing: 8.7 m
Max range: ERFB-BB 39000 m
Muzzle velocity: 897 m/s
Projectile weight: ERFB-BB HE 47.6 kg
Depression/elevation: -3°/+75°
Traverse: 65°

Bofors 155 mm Field Howitzer FH-77B — Sweden

In 1973 Bofors AB produced the prototype of a 155 mm howitzer intended for the Swedish Army. By 1978 the type was placed in production as the FH-77A. As the FH-77A uses an ammunition system unique to the Swedish Army, one not likely to appeal to the international market, Bofors produced a version capable of using NATO standard 155 mm ammunition. This became the **FH-77B**, subsequently sold to India in one large 410 unit order, Nigeria and the Swedish Army.

The **FH-77B** is essentially similar to the FH-77A but the 39-calibre ordnance has a chamber and rifling intended for standard NATO ammunition systems. Early barrels had a pepper-pot muzzle brake but later versions use a single-baffle brake. The **FH-77B** carriage has many advanced features, not the least of which are powered systems intended to minimise crew fatigue. An auxiliary power unit (APU) mounted on the front carriage provides hydraulic power to open and close the trail legs, raise and lower the trail dolly wheels, provide power to move the **FH-77B** over short distances, power an ammunition crane, the loading system and a power rammer, and provide power to lay the barrel. It is theoretically possible for only one or two men to deploy and serve the gun in action; the full crew is six.

On the move a tractor vehicle can engage the APU to provide extra power when traversing difficult terrain. In action the **FH-77B** is fired from its wheels. Projectiles are raised from pallets in 'clips' of three projectiles which are lowered into the powered loading system; all three projectiles can be fired within 12 seconds with bagged propellant charges being hand-loaded.

The **FH-77B** can fire all 155 mm NATO standard and ERFB projectiles although Bofors produce their own HE and HE extended range projectiles. Standard Bofors projectiles can reach 24000 metres and the extended range versions, 30000 metres.

A mobile coast defence version has been proposed.

155 mm FH-77B

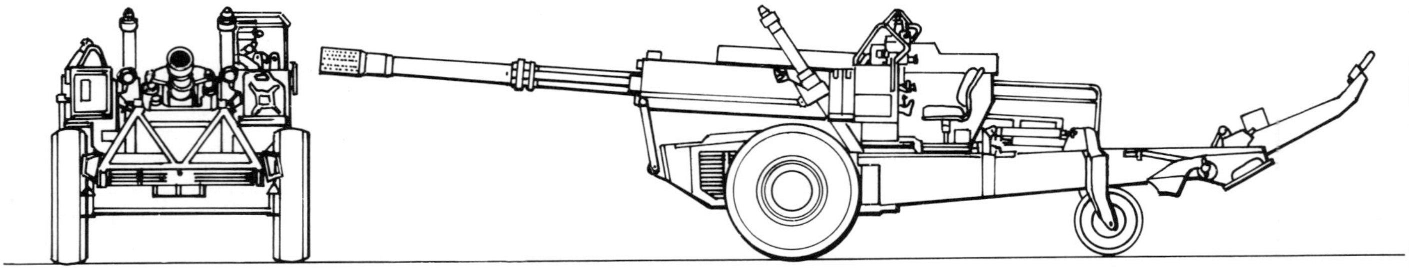

Specification

First prototype: 1973
First production: 1981
Current users: India, Nigeria and Sweden
Crew: 6
Weight in action: 12000 kg
Barrel length: 6.045 m
Length firing: 11.16 m
Width firing: 7.18 m
Max range: standard HE 24000 m;
extended range HE 30000 m
Muzzle velocity: 827 m/s
Projectile weight: HE 42.6 kg
Depression/elevation: -3°/+70°
Traverse: 60°

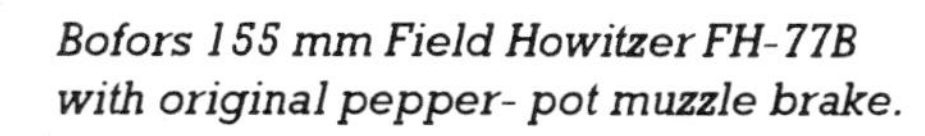

Bofors 155 mm Field Howitzer FH-77B with original pepper- pot muzzle brake.

155 mm Field Howitzer 70 (FH-70) — International

The **155 mm FH-70** howitzer is a rare example of an international defence project coming to a successful fruition, being a programme carried out by the United Kingdom, Germany and Italy to meet a common requirement. Development, which began as far back as the early 1960s, was protracted with no fewer than 19 pre-production prototypes being manufactured before production commenced in 1978. Production was allocated between the three participants and a new family of 155 mm ammunition was developed for the **FH-70**.

The **FH-70** is typical of its design generation, being a large and rather heavy howitzer with a 39-calibre barrel. There is a large double-baffle muzzle brake and a vertical sliding block breech. The ordnance is mounted on a complex split trail carriage with an auxiliary power unit (APU) at the front to power the carriage over short distances and provide hydraulic power for services such as opening and closing the trail legs. The hydraulic system is also used to raise the trail legs for large changes in traverse. On tow the ordnance is reversed over the trails which are supported on two dolly wheels. In action a semi-automatic loading system enables three rounds to be fired in 13 seconds.

The ammunition system developed for **FH-70** is based around a new L15 HE projectile and a bagged propellant system, to be replaced by a five-part 'unicharge' system using equal increments in rigid combustible containers. The standard L15 HE shell has a range of 24700 metres at full charge while enhanced range base bleed (BB) projectiles can reach 31500 metres.

Production of **FH-70** has ceased in Europe but licence production continues in Japan for the Self-defence Forces. Exports were made to Saudi Arabia and Malaysia.

Rheinmetall of Germany produced a **FH-70** prototype with a 52-calibre barrel. This version has a range of 30000 metres firing the L15 HE projectile.

155 mm FH- 70

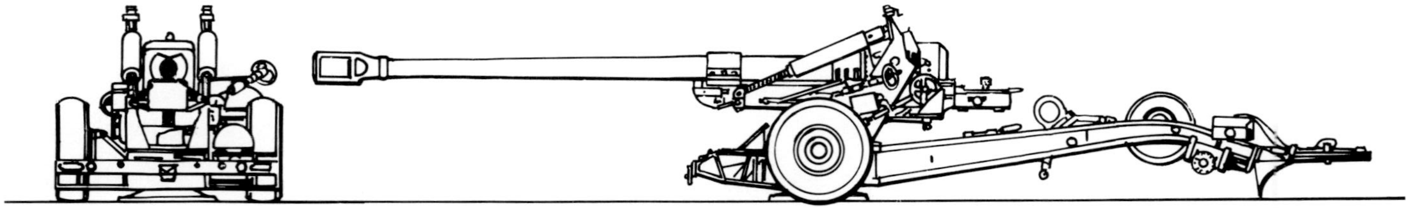

Specification

First prototype: 1969
First production: 1978
Current users: United Kingdom, Germany, Italy, Saudi Arabia, Malaysia, Japan
Crew: 7 or 8
Weight in action: 9300 kg
Barrel length: 6.022 m
Length firing: 12.43 m
Width firing: 7.5 m
Max range: L16 HE 24700 m; BB 31500 m
Muzzle velocity: 827 m/s
Projectile weight: 42.55 kg
Depression/elevation: -4.5°/+70°
Traverse: 56°

A battery of 155 mm FH-70s on the Larkhill ranges.

Vickers 155 mm Ultralightweight Field Howitzer

UK

The **Vickers 155 mm Ultralightweight Field Howitzer (UFH)** was developed in response to a US Army requirement for a towed 155 mm howitzer weighing no more than 4000 kg - Royal Ordnance 155 mm Light Towed Howitzer (see following entry) was developed to meet the same requirement. Vickers produced the first of two **UFH** prototypes in 1989 and in 1993 they were still undergoing evaluation and other trials for the US Marine Corps as well as the US Army. The British Army may procure whichever design emerges successfully from the trials.

In order to keep weight down sufficiently for the **UFH** to be carried slung under a UH-60 Blackhawk helicopter Vickers Shipbuilding and Engineering Limited (VSEL) introduced some novel design techniques for the **UFH**, including the extensive use of titanium alloys and other advanced materials to reduce weight. The carriage is of the leading arm type with two trail legs laying forward of the firing platform - two trails at the rear act as further stabilisers. The 39-calibre barrel, which is essentially the same as that mounted on the M109A6 Paladin, has a large double-baffle muzzle brake with a towing eye as the **UFH** is towed by the barrel, with the forward trails folded backwards and the rear legs folded upwards - two standard truck wheels are lowered for moves, one each side of the firing platform. Time in and out of action is under two minutes. If required, it is possible to break the **UFH** down into two main loads. The **UFH** is air portable in one load.

As the **UFH** is intended to have the same ballistic performance as the in-service 155 mm M198 towed howitzers, it fires all standard NATO ammunition to a maximum range of 24700 metres - enhanced range base bleed (BB) projectiles can reach 30000 metres. As the **UFH** does not have an auto-loading system (although one is under development) the rate of fire is around four rounds a minute.

155 mm UFH

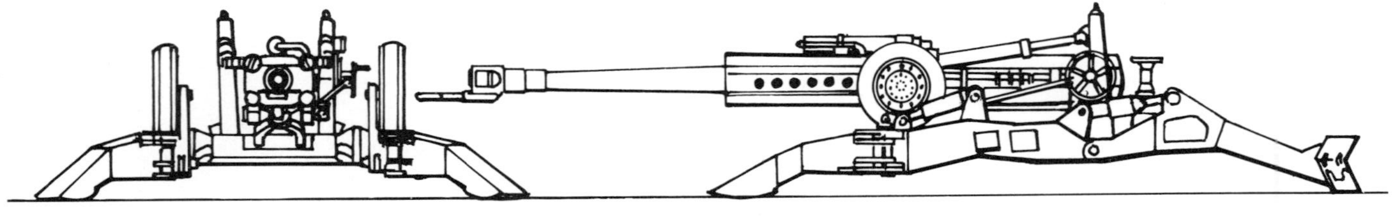

Specification

First prototype: 1989
First production: not yet in production
Current user: not yet in production
Crew: 6 to 8
Weight in action: 3745 kg
Barrel length: 6.096 m
Length firing: 10.21 m

Width firing: 3.72 m
Max range: standard 24700 m;
BB 30000 m
Muzzle velocity: 827 m/s
Projectile weight: 43.35 kg
Depression/elevation: -5°/+70°
Traverse: 45°

Vickers 155 mm Ultralightweight Field Howitzer on firing range.

Royal Ordnance
155 mm Light Towed Howitzer — UK

The Royal Ordnance **155 mm Light Towed Howitzer (LTH)** was developed to meet the same US Army requirement as the VSEL UFH (see previous entry) with the first of two prototypes being ready for trials in early 1991. The **LTH** has undergone the same series of evaluation and other trials as the UFH although Royal Ordnance regard their **LTH** as the prime market replacement for their 105 mm Light Gun (qv) as well as the possible candidate for the US armed forces requirement. As a result several versions of the **LTH** are in the planning stage.

Royal Ordnance use the same ordnance as the UFH, ie a modified version of that used on the self-propelled M109A6 Paladin although the **LTH** involves a more conventional carriage compared to the UFH. Extensive use is made of titanium alloys and other advanced materials, including aluminium alloys for the upper carriage. The **LTH** carriage is of conventional split trail form even though the barrel is traversed over the trails for towing. The recoil system is unusual as the Royal Ordnance designers altered the usual straight recoil movement to a curvilinear system in which the recoiling barrel moves in an upwards arc, thereby saving both weight and length of recoil.

As the **LTH** is intended to have the same ballistic performance as the in-service 155 mm M198 towed howitzers, it fires all standard NATO ammunition to a maximum range of 24700 metres - enhanced range base bleed (BB) projectiles can reach 30000 metres.

Royal Ordnance are proposing a series of **LTH** variants with various barrel lengths to meet virtually any possible weight and range requirement. As well as the 39-calibre barrel used on the prototypes it is intended that the **LTH** carriage, possibly with folding trail legs, could accommodate 25, 35, 45 or 52-calibre barrels, with a high degree of interchangeability between all the variants.

155 mm LTH

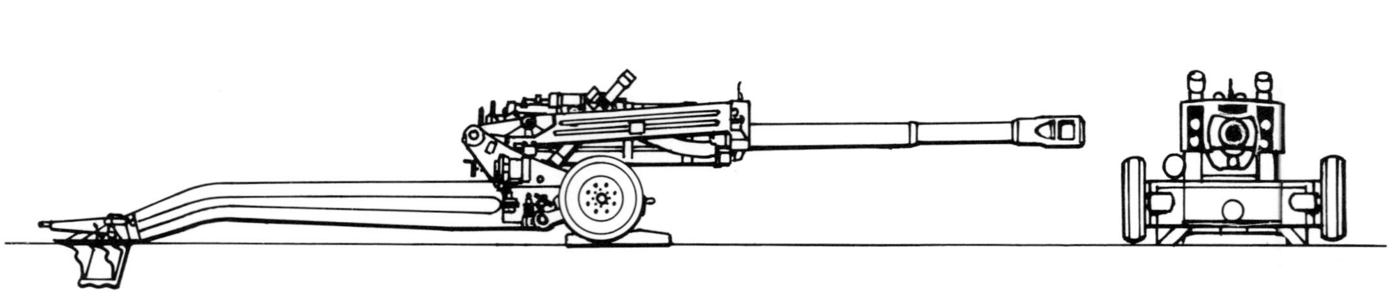

Specification

First prototype: 1991
First production: not yet in production
Current user: not yet in production
Crew: 6 to 8
Weight in action: 4060 kg
Barrel length: 6.096 m
Length firing: 11.938 m
Width firing: 7.5 m
Max range: standard 24700 m;
BB 30000 m
Muzzle velocity: 827 m/s
Projectile weight: 43.5 kg
Depression/elevation: -5°/+72°
Traverse: 45°

Royal Ordnance 155 mm Light Towed Howitzer at maximum elevation.

Royal Ordnance 105 mm Light Gun UK

The Royal Ordnance **105 mm Light Gun** was developed from 1965 onwards to replace the OTO Melara Model 56 pack howitzers (qv) then in service with the Royal Artillery and has been a major sales success for Royal Ordnance, achieving a breakthrough into the difficult American defence market with substantial orders for the US Army who have a requirement for nearly 600 units.

The standard **Light Gun** has an unusual bow-shaped carriage with tubular trails arranged so that the wheels can rest on a circular firing platform which enables rapid changes in traverse to be carried out by one man; the lightweight platform is carried over the trails during moves. The ordnance rests in a saddle but due to its high centre of axis the 30-calibre barrel has to be reversed to over the trail for towing and when coming into action. This necessitates the removal of one road wheel, a process rapidly carried out using a small jack and a knock-off hub cab.

The standard British Army **Light Gun** is the L118 which fires a special family of 105 mm separate loading ammunition originally developed for use with the Abbot self-propelled gun. (The **Light Gun** ordnance was developed using the Abbot barrel as the basis - the Abbot has now been withdrawn from service). However a revised barrel can be installed to fire the readily-available American/NATO 105 mm ammunition based on the HE M1. This model is known as the L119, the version the US Army has adopted as the M119. The M119 has a maximum range of 11500 metres compared to the 17200 metres possible with the L119.

American M119s are licence-produced in the United States. Other licence production was carried out in Australia for both the Australian and New Zealand armies. India Ordnance Factories have developed a close copy of the Light Gun known as the 105 mm Light Field Gun.

105 mm Light Gun

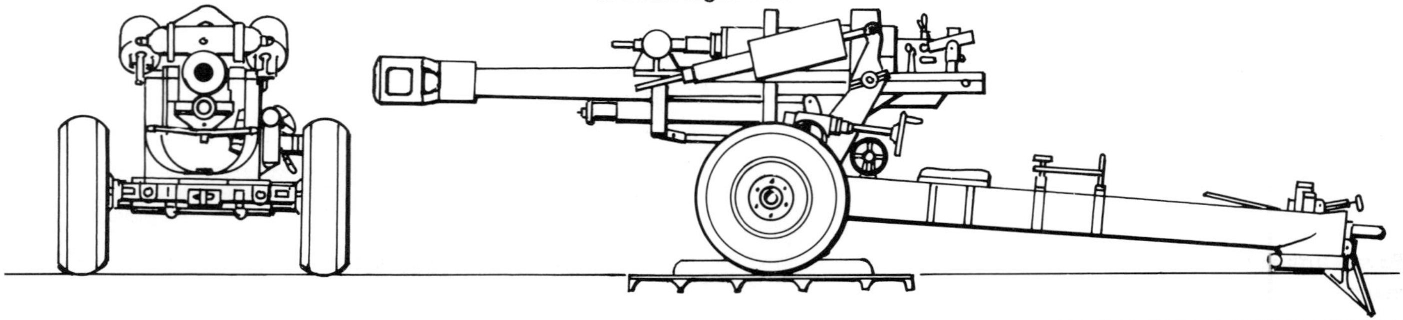

Specification

First prototype: late 1960s
First production: 1974
Current users: UK, USA and at least 12 other nations
Crew: 6
Weight in action: 1860 kg
Barrel length: 3.4 m
Length firing: 7.01 m
Width firing: 1.778 m
Max range: 17200 m
Muzzle velocity: 708 m/s
Projectile weight: 16 kg
Depression/elevation: -5.5°/+70°
Traverse: on carriage 11°;
on platform 360°

Royal Ordnance 105 mm Light Gun on manoeuvres.

105 mm Howitzer M101

The **105 mm Howitzer M101** is one of the veterans of the artillery world having been first mooted in 1919. The first examples, then known as the 105 mm Howitzer M1, were produced in 1928 and thereafter the M1 became the M2, redesignated the **M101** after 1945. Since 1940, when the M2 was first produced in significant numbers, the **M101** series has been produced in thousands, with the main production run ceasing in 1953, only to be re-started at intervals to meet special orders.

In design terms the **M101** is entirely conventional, with a split trail carriage and a general standard of construction which can only be described as sturdy. The barrel lacks a muzzle brake (one is fitted to Belgian and German examples) and is approximately 22 calibres long with the recoil mechanisms located over and below the barrel itself. A shield is normally fitted.

The **M101** has come to be regarded as the standard by which other pieces are compared. This is particularly true regarding the ammunition fired from the **M101** and especially the HE M1, probably the most widely-used 105 mm projectile in service. The HE M1 has been joined by a wide array of other projectiles, including cargo rounds, anti-personnel flechette containers, smoke, illuminating and various anti-armour projectiles. Ammunition in the series is produced all around the world.

The main drawback of the **M101** for many armies is the maximum range of only 11270 metres. To overcome this, some manufacturers have introduced rocket-assisted (RAP) or base bleed (BB) projectiles with increased ranges, while others have proposed that new longer 33-calibre barrels could replace the existing barrels to provide a range increase to 15700 metres or even around 20000 metres with enhanced range projectiles. Conversions are on offer from companies in Germany, France and the Netherlands. The South Korean KH178 (qv) is based on the **M101** carriage.

The US Army is replacing its **M101s** with the M119 Light Gun (see previous entry) but many other nations will continue to retain their **M101s** for decades to come.

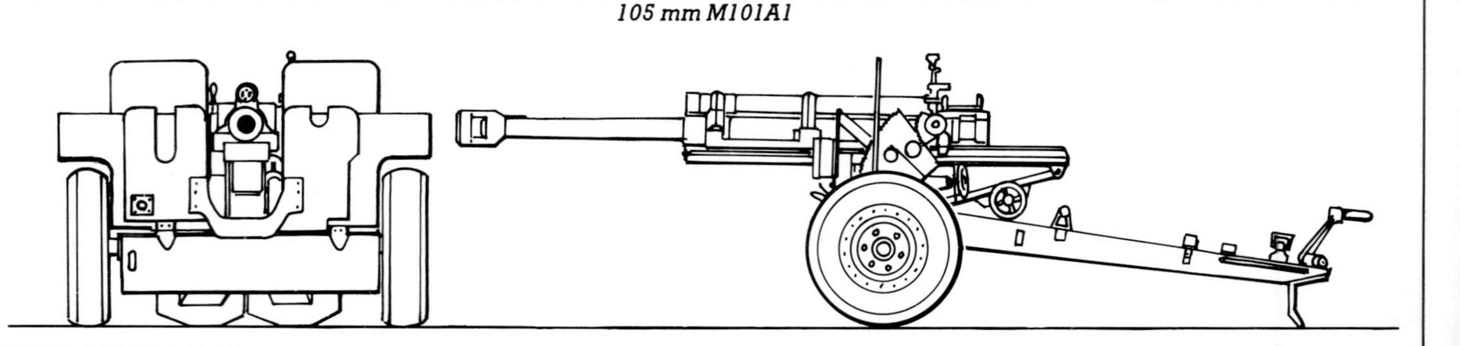

105 mm M101A1

Specification

First prototype: 1925
First production: 1940
Current users: well over 60 countries
Crew: 8
Weight in action: 2030 kg
Barrel length: 2.363 m
Length firing: 5.99 m
Width firing: 3.657 m
Max range: HE M1 11270 m
Muzzle velocity: 472.4 m/s
Projectile weight: 14.97 kg
Depression/elevation: -5°/+66°
Traverse: 46°

US Army 105 mm Howitzer M101 on the ranges.

105 mm Howitzer M102

USA

In 1960 the US Army decided to develop a new 105 mm howitzer to replace their existing M101s (see previous entry). A prototype was duly produced in 1962 and type classified in late 1963 as the **105 mm Howitzer M102**. The main attraction of the **M102** was that it weighed much less than the M101 but set against this the range performance was only a slight improvement over the older model. This was one reason why the **M102** was not a great success with the US Army, especially after major design modifications had to be introduced to overcome problems experienced in Vietnam. Production ceased in 1970 after only five years - even by then the **M102** had only supplemented rather than replaced the M101. The US Army's **M102** are due to be replaced by the 105 mm Light Gun M119 (qv).

The **M102** has a box trail carriage constructed using aluminium and resting on a firing baseplate which permits rapid and easy changes of traverse through a full 360°. The barrel is slightly longer than that of the M101 and slides in an assembly which contains the recoil mechanisms. A vertical sliding breech block is used. There is no muzzle brake.

The **M102** fires the same large family of ammunition as the M101, with the maximum range firing the HE M1 being 11500 metres. To improve the range potential a rocket-assisted projectile (RAP), known as the HERA M548, was introduced to increase the maximum range to 15100 metres, but with a loss in accuracy at the longer ranges. Another rocket-assisted enhanced range projectile, known as the HERA XM927, can reach 16500 metres but it is not yet in service. Cargo projectiles containing grenades were developed for use in Vietnam.

For most of its US Army service life the **M102** has tended to be used mainly by high mobility forces and airborne units but those which were exported, often as military aid, are often issued to field units.

105 mm M102

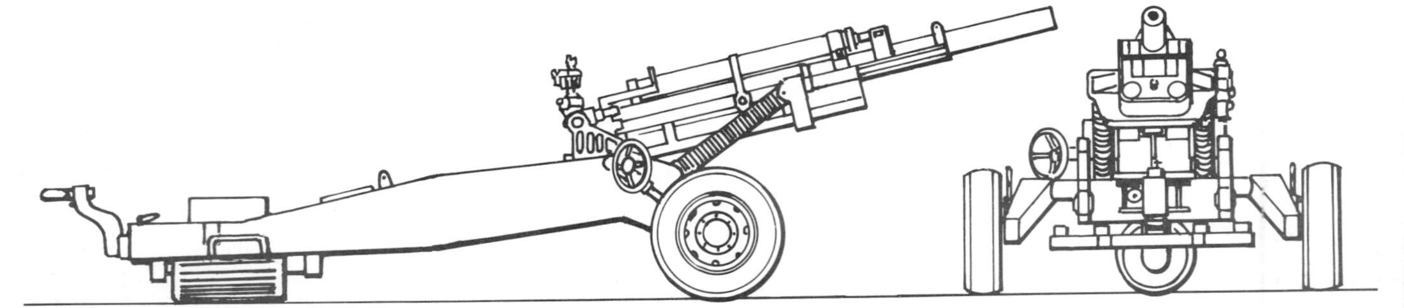

Specification
First prototype: 1962
First production: 1965
Current user: USA, France, Greece, Israel, Saudi Arabia and at least 15 other countries
Crew: 8
Weight in action: 1496 kg
Barrel length: (overall) 3.382 m
Length firing: 5.182 m
Width firing: 1.964 m
Max range: HE M1 11500 m;
HERA M548 15100 m
Muzzle velocity: HE M1 494 m/s
Projectile weight: 14.97 kg
Depression/elevation: -5°/+75°
Traverse: 360°

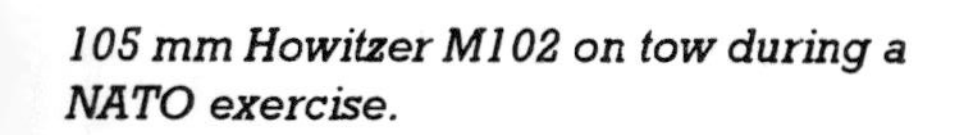

105 mm Howitzer M102 on tow during a NATO exercise.

155 mm Howitzer M114

The **155 mm Howitzer M114** is another long-lived American field piece which can trace its origins to before the Second World War when the US Army used 155 mm howitzers of French extraction. The US Army updated the old designs for mechanised traction and added split trails to produce the 155 mm Howitzer M1, redesignated the **M114** after 1945. Production commenced in 1941 - by the time it ended over 6000 had been made. Slight carriage changes later produced the M114A1. Licence production of the **M114** is still undertaken in South Korea.

The **M114** is a strong design which is basically simple. The split trail carriage supports the upper carriage on which the 23-calibre barrel is mounted - there is no muzzle brake. The basic carriage is considered strong enough to permit it to be converted to accept 39-calibre barrels. One US Army project was to place the ordnance from the M198 howitzer (see following entry) on the **M114** carriage to produce the M114A2 but not many M114A2 conversions appear to have been made. Outside the USA several commercial concerns have proposed similar conversions, including RDM of the Netherlands (qv), Giat of France (the M114F), OTO Melara of Italy and TAAS - Israel Industries. The South Korean KH179 (qv) is based on the **M114** while the former Yugoslavia produced numbers of a **M114** clone known as the M65. Taiwan has mounted numbers of **M114s** on locally converted open-topped M108 or M109 self-propelled carriages - these are known as the XT-69.

The **M114** fires a large array of ammunition natures, one of which is numerically the most important of all 155 mm projectiles, the HE M107. The HE M107 is manufactured by numerous concerns all around the world (even in China) and is still regarded as the 'standard' 155 mm projectile, despite its indifferent range and fragmentation performance when compared to later designs. Other projectiles in the **M114** ammunition family include smoke and illuminating but the inability of the **M114** to utilise the more powerful propellant charges developed for the longer-barrelled howitzers limits the **M114's** range to around 14600 metres.

155 mm M114

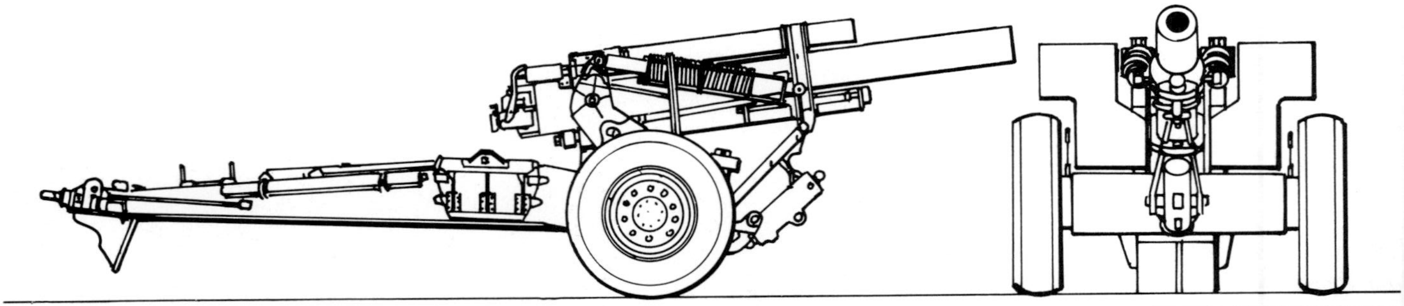

Specification

First prototype: 1940
First production: 1941
Current users: USA and well over 40 other countries
Crew: 11
Weight in action: 5760 kg
Barrel length: 3.626 m
Length travelling: 7.315 m
Width travelling: 2.438 m
Max range: 14600 m
Muzzle velocity: 564 m/s
Projectile weight: 43.88 kg
Depression/elevation: -2°/+63°
Traverse: 49°

155 mm Howitzer M114 on tow by high speed tracked tractor.

155 mm Howitzer M198

In 1968 the US Army requested a 155 mm howitzer to replace their ageing M114s (see previous entry). Development of a design known as the XM198 was carried out by the Rock Island Arsenal with the first of a series of prototypes appearing in 1969. Production of the type classified version, the **M198**, commenced in 1978. Since then over 1700 **M198s** have been manufactured, over 1300 for the US Army and Marine Corps alone, and production continues.

The **M198** gives a visual impression of being heavy and awkward, factors highlighted by the absence of the usual auxiliary power unit (APU), although one was developed. However, most users, including the US armed forces, seem content with their **M198s** and it performed well during the 1991 Gulf operations.

Overall, the **M198s** design resembles others of its generation, including split trails supporting an upper carriage carrying a 39-calibre barrel. The barrel, which has a double-baffle muzzle brake, is ballistically matched with other NATO standard ordnance such as the FH-70 (qv) so it can fire all the latest types of ammunition, including ERFB natures although these are not held by the US armed forces.

The HE M107 is still fired although the maximum range is limited to 18150 metres. More advanced types such as the M483A1 ICM (improved conventional munition, also known as a cargo round), carrying 88 dual-purpose bomblets, can reach 22000 metres. The rocket-assisted HERA M549A1 has a maximum possible range of 30100 metres.

Other ammunition types fired by the **M198** include the M718/M741 cargo projectiles which can create anti-tank or anti-personnel minefields at long ranges, and the M712 Copperhead, a laser-guided anti-armour projectile with a maximum range of 16000 metres, although operational ranges are usually much shorter. Also available for the **M198** is a new range of smoke, illuminating and radio jamming projectiles based on the M483A1 ICM projectile body. The M454 nuclear projectile has been withdrawn.

155 mm M198

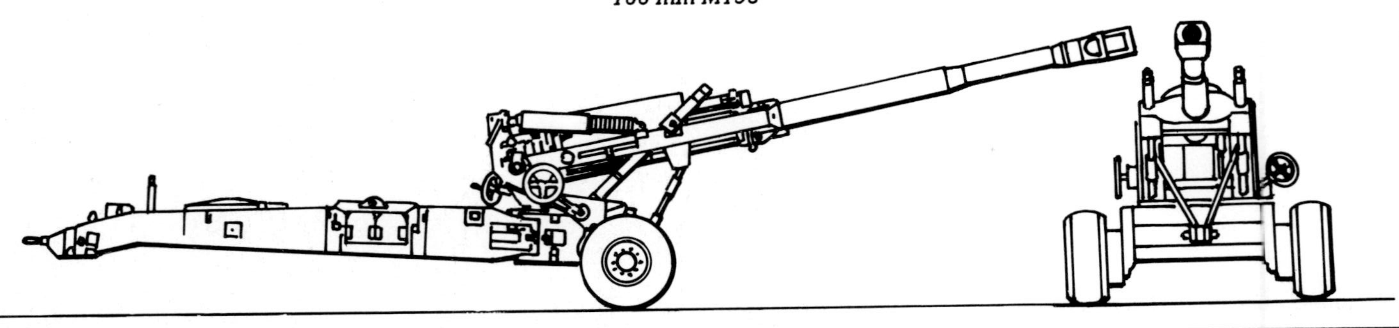

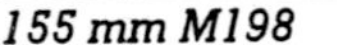

Specification
First prototype: 1969
First production: 1978
Current users: USA, Pakistan, Australia, Bahrain, Saudi Arabia, Thailand, Tunisia and others.
Crew: 11
Weight in action: 7163 kg
Barrel length: 6.096 m
Length firing: 11 m
Width firing: 8.534 m
Max range: M483A1 ICM 22000 m; HERA M549A1 30100 m
Muzzle velocity: HE M107 684 m/s
Projectile weight: 43.88 kg
Depression/elevation: -5°/+72°
Traverse: 45°

155 mm Howitzer M198 in action.

Type 83 152 mm Self-propelled Gun-howitzer

China

The **Type 83 152 mm Self-propelled Gun-howitzer** is produced by China North Industries Corporation (NORINCO) and although it was first observed during 1983 there are still some doubts as to when it was first produced. The overall design of the **Type 83** is 'classic' for a self-propelled artillery piece as the main gun turret is located towards the rear of the hull, with the engine compartment at the front and the driver seated to the hull left. A 12.7 mm machine gun is carried on the hull roof for air and local defence while next to the main barrel is a co-axial 7.62 mm machine gun.

The 152 mm howitzer carried by the **Type 83** is a variant of the towed 152 mm Howitzer Type 66, the Chinese-produced version of the CIS gun-howitzer D-20 (qv). It is understood that the **Type 83** fires the same ammunition as the towed Type 66 which, for the Chinese armed forces, is limited to two main types, HE and smoke, both with a maximum range of 17230 metres. There have been reports of a Chinese-developed rocket-assisted projectile (RAP) which can reach 21880 metres with, according to Chinese claims, no loss in accuracy at the longer ranges as is usually the case with RAPs. The **Type 83** has the capacity to carry 30 projectiles and charges, with re-loading being accomplished through a door in the hull rear or via doors in the turret sides and rear.

The chassis used for the **Type 83** is of a special type. Since the **Type 83** was introduced the same chassis has been modified to carry a trench digging machine, a mine clearing rocket system and the **Type 83** 122 mm self-propelled multiple rocket system (qv) with 40 barrels and an automatic reloader. Although production of the **Type 83** has now ceased, production of the basic chassis may continue for the variants or for some as yet undisclosed use(s).

152 mm Type 83

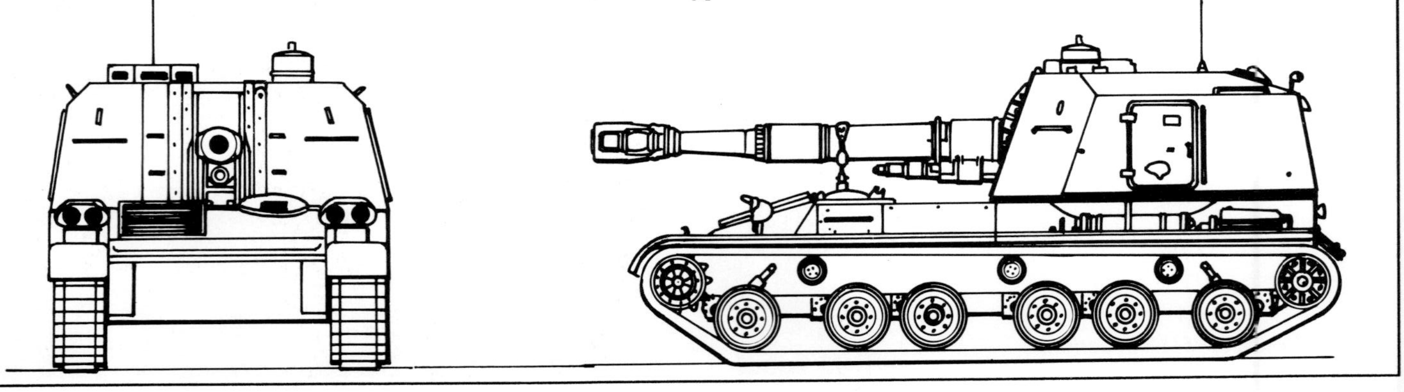

Specification

First prototype: not known
First production: possibly 1982
Current user: China
Crew: 4 or 5
Weight in action: 30000 kg
Length overall: 7.33 m
Length of hull: 6.882 m
Width: 3.236 m
Height: 3.502 m
Road range: 450 km
Fording: 1.3 m
Powerpack: Type 12150L diesel developing 520 hp
Depression/elevation: -5°/+65°
Traverse: 360°

Type 83 152 mm Self-propelled Gun-howitzer prepared for parade.

120 mm SO-120
Self-propelled Howitzer/Mortar

The **120 mm SO-120**, or **2S9 Anona (Anemone)**, is a hybrid weapon system combining the attributes of the indirect fire mortar with the direct fire capabilities of the gun-howitzer. It was developed to provide the CIS airborne forces with a light multi-purpose artillery support system. It is understood that over 1000 **SO-120s** were produced.

The **SO-120** resembles a light tank but the ordnance carried is a 120 mm breech-loaded rifled mortar, hand-loaded by one of the two men in the turret, the other being the gunner. Up to 60 rounds can be carried. The vehicle commander is not in the turret but in a position in the front left of the hull where he is provided with vision devices. The driver doubles as a mechanic. To provide full mobility the **SO-120** is amphibious and can be para-dropped.

The 120 mm mortar has a rate of fire of six to eight rounds a minute. When firing the vehicle suspension is lowered. In theory the mortar can fire most types of 120 mm mortar ammunition produced in either the East or West but special rounds are produced specifically for the **SO-120**, based around a boat-tailed HE projectile weighing 19.8 kg. When fired, with a muzzle velocity of 367 m/s, the propellant charge is fixed to the projectile base; maximum range is 8855 metres. Maximum range can be increased by the use of a rocket-assisted HE round (HE-RAP) which has a range of 13000 metres.

For direct fire against armoured and other targets the **SO-120** fires a fin-stabilised high explosive anti-tank (HEAT) projectile weighing 13.17 kg and having a direct fire range of about 500 metres. Armour penetration is about 600 mm.

To emphasise its artillery role the **SO-120** does not have a roof-mounted or co-axial machine gun.

The turret and mortar/howitzer of the **SO-120** is also used on the 120 mm 2S23, a modified BTR-80 8 x 8 armoured personnel carrier.

120 mm SO-120

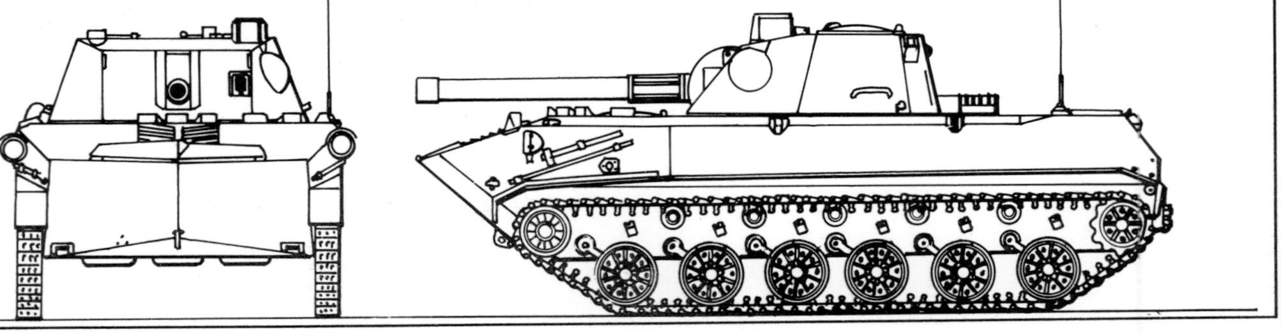

Specification

First prototype: early 1980s
First production: 1984(?)
Current users: CIS and Afghanistan
Crew: 4
Weight in action: 8700 kg
Length overall: 6.02 m
Length of hull: 6.02 m
Width: 2.63 m
Height: travelling 2.3 m; firing 1.9 m
Road range: 500 km
Fording: amphibious
Powerpack: 5D20 diesel developing 240 hp
Depression/elevation: -4°/+80°
Traverse: 70°

120 mm SO-120 Self-propelled Howitzer/Mortar parading in Moscow.

122 mm SO-122 Self-propelled Howitzer (2S1)

Former Soviet Union

Developed during the late 1960s the **122 mm SO-122 self-propelled howitzer** entered service in 1972 but it was not until 1974 that it was first noted by Western observers, hence the Western designation M-1974. The **SO-122** is also known as the 2S1 or Gvozdika (Carnation). It is based on a lengthened MT-LB tracked multi-purpose carrier chassis and carries a modified version of the ordnance used by the towed 122 mm D-30 howitzer (qv).

The **SO-122** is numerically one of the most important of all CIS armoured support vehicles (over 10000 were produced) having been utilised for all manner of purposes apart from the self-propelled howitzer. In the latter role the **SO-122** uses the 'classic' rear-mounted turret having a flat profile and housing the ordnance and three of the four-man crew - more crew members are carried in an accompanying ammunition carrier. The ordnance, known as the 2A31, uses a fume extractor, a double-baffle muzzle brake, and a vertical sliding block breech mechanism. A power rammer and extractor are fitted to provide a sustained rate of fire of between five and eight rounds a minute.

The **SO-122** normally carries 40 projectiles and charges, although this is reduced to 30 when the vehicle is required to utilise its amphibious capabilities. For most fire missions the types are limited to HE, smoke and a few fin-stabilised HEAT anti-armour projectiles; the normal HE range is 15300 metres. A rocket-assisted projectile (RAP) with a maximum range of 21900 metres is available, as are leaflet, anti-personnel flechette carriers and illuminating; a chemical projectile has been withdrawn.

The basic hull of the **SO-122** is used for a number of associated turretless battery fire control and artillery reconnaissance carriers as well as for battlefield surveillance radar carriers, NBC reconnaissance vehicles, mineclearing vehicles, and command vehicles (among others). Production of the **SO-122** ceased in 1991 although conversions of the basic chassis continue.

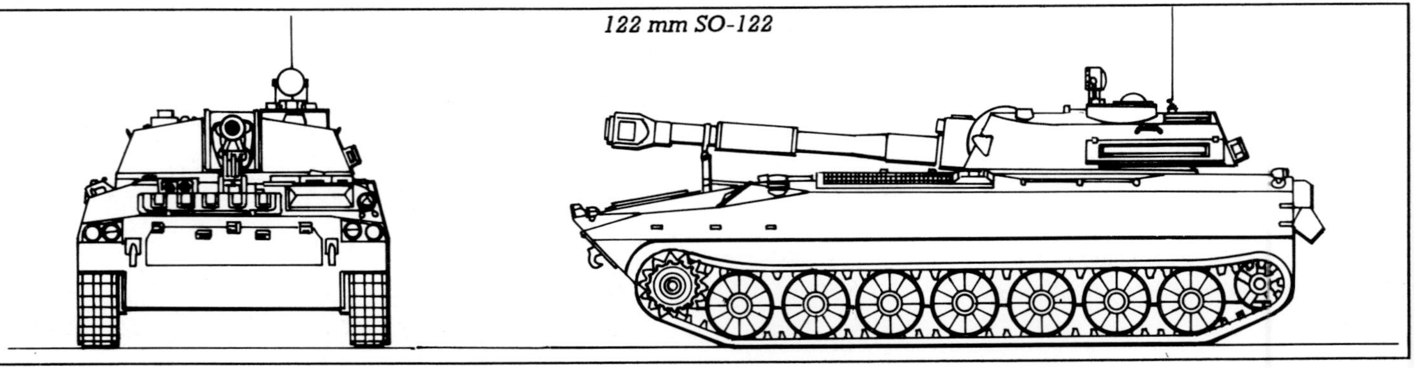

122 mm SO-122

Specification

First prototype: late 1960s
First production: 1981-1982
Current users: CIS, Algeria, Angola, Bulgaria, Czech & Slovak Republics, Hungary, Iraq, Libya, Poland, Syria, former Yugoslavia
Crew: 4
Weight in action: 15700 kg
Length overall: 7.26 m
Length of hull: 7.26 m
Width: 2.85 m
Height: 2.725 m
Road range: 500 km
Fording: amphibious
Powerpack: YaMZ-238v diesel developing 240 hp
Depression/elevation: -3°/+70°
Traverse: 360°

122 mm SO-122 Self-propelled Howitzer (2S1) with commander and driver accesses open.

130 mm Coast Defence Mobile Gun System

Former Soviet Union

The current status of this **130 mm coast defence gun system** is uncertain, and not even its correct designation is known, but it is an interesting development intended to provide defensive cover against amphibious operations along exposed lengths of coastline.

First revealed in 1993, the system is based around a 130 mm naval gun mounted in a turret carried by a variant of the massive MAZ-543 8 x 8 high mobility truck.

The type of **130 mm gun** is uncertain but it has a 60 to 70-calibre barrel with a muzzle brake and fume extractor. Mounted in a traversing turret over the rear axles, the gun is normally carried with the barrel pointed forwards over the driver's and commander's side-mounted cab. The rest of the eight-man crew travel in a compartment behind the cab which also contains some of the 40-round ammunition load plus supplies to maintain the vehicle in the field for prolonged periods. On arrival at a firing site the gun can be in action within minutes and thereafter can fire a minimum of ten rounds a minute against sea targets, including surface-effect craft travelling at up to 200 knots. When in action the vehicle is raised on hydraulic jacks. Each gun has ammunition handling systems to maintain the fire rate. Projectile weight is probably around 27 kg and may include armour-piercing as well as the usual HE.

The **130 mm gun** vehicle normally operates under the direct control of a central multi-sensor fire command and control vehicle, also a converted MAZ-543, controlling up to six guns - associated with the control vehicle is a further support vehicle providing electrical power and other support functions. In action a battery has a maximum operating range of 20000 metres although maximum range is 27000 metres. If required individual vehicles can operate autonomously as each has its own ballistic computer, an optronics fire control system and rangefinder, plus an on-board power generator.

130 mm CDMGS

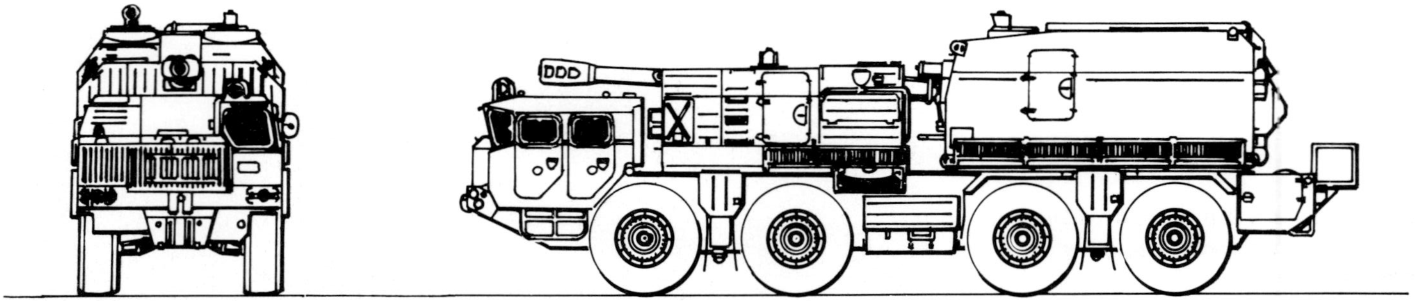

Specification

First prototype: late 1980s
First production: exact production status uncertain
Current user: uncertain
Crew: 8
Weight in action: 43700 kg
Length travelling: 12.95 m
Width: 3.1 m
Height travelling: 3.925 m
Road range: 650 km
Fording: 1.2 m
Powerpack: D12A-525 V-12 diesel developing 525 hp
Depression/elevation: -5°/+50°
Traverse: 240°

An emplaced 130 mm Coast Defence Mobile Gun System at maximum elevation.

152 mm SO-152
Self-propelled Gun-howitzer

Former Soviet Union

The **152 mm SO-152 self-propelled gun-howitzer** may be regarded as a parallel development to the SO-122 having been developed during the late 1960s to enter production in 1971. It was not until 1973 that the **SO-152** was first noted by the West, hence their designation of M-1973; other names are **2S3** and Akatsiya (Acacia).

The **SO-152** uses its own specially-developed chassis with the turret at the rear and mounting a modified version (the 2A33) of the 152 mm howitzer D-20 ordnance, (qv) but with the addition of a fume extractor located behind the double-baffle muzzle brake. Outwardly the **SO-152** resembles the American M109 series but there are many differences. For instance, the **SO-152** is not amphibious.

As the **SO-152** fires the same separate-loading ammunition as the towed D-20 the ranges are the same, ie a 43.5 kg HE projectile can be fired to a maximum range of 18500 metres. Inside the turret 12 projectiles are loaded in a rotary magazine which presents the selected projectiles and charges to an autoloader system; early models lacked this refinement. The usual procedure is for ammunition to be loaded, round-by-round, through circular doors in the hull rear, directly from a truck parked close to the rear of the **SO-152**. In this way the full load of 35 projectiles in the turret and hull can be conserved for use when immediate resupply is not possible. (The load is 40 on the later models with the rotary magazine, recognisable externally by a single large rear hull door).

Apart from the standard HE rounds the **SO-152** can fire the laser-guided Krasnopol anti-armour missile, a fin-stabilised HEAT anti-armour projectile as well as the usual illuminating, smoke and leaflet projectiles. Chemical and tactical nuclear projectiles have been withdrawn. A rocket-assisted projectile (RAP) with a maximum range of 24000 metres is also available.

152 mm SO-152

Specification

First prototype: late 1960s
First production: 1971
Current users: CIS, Hungary, Iraq, Libya, Syria.
Crew: 4
Weight in action: 27500 m
Length overall: 8.4 m
Length of hull: 7.765 m
Width: 3.25 m
Height: 3.05 m
Road range: 500 km
Fording: 1 m
Powerpack: V-59 diesel developing 520 hp
Depression/elevation: -4°/+60°
Traverse: 360°

152 mm SO-152 Self-propelled Gun-howitzer in travelling lock.

152 mm 2S5 Self-propelled Gun Former Soviet Union

During the mid 1970s the old Soviet Union developed two 152 mm self-propelled artillery systems. One became the towed 2A36 gun (qv) while the other became the **2S5 self-propelled gun**. Development of the **2S5** paralleled that of the towed gun and it was 1976 when the first examples were produced. The **2S5** has never been exported.

The **152 mm 2S5 self-propelled gun** uses an open configuration, no doubt as the long ranges at which the **2S5** are expected to be used places them away from most battlefield retaliatory hazards. The chassis on which the ordnance is mounted bears similarities with that of the 152 mm 2S3 (qv) but on the **2S5** the 49-calibre ordnance is mounted on the rear of the open chassis with only a small shield provided to protect the gun systems and crew. At the rear of the chassis is a large spade assembly which is lowered to provide stability when firing. A dozer blade is provided at the hull front to prepare firing positions and remove battlefield obstacles.

The multi-baffle muzzle-braked barrel of the **2S5** gun produces a high muzzle velocity (approximately 800 m/s) and a correspondingly long range of 27000 metres firing standard projectiles (some references state 28400 metres) and a reported 40000 metres with a rocket-assisted projectile (RAP). The **2S5** fires a family of streamlined separate-loading ammunition shared only by the self-propelled 2A36 including, at one period, a tactical nuclear projectile. Other rounds include a FRAG-HE (the most widely used, weighing 46 kg), concrete-piercing, incendiary, and chemical (now withdrawn). For direct fire use against armoured targets the **2S5** fires an armour-piercing round. The Krasnopol laser-guided anti-armour projectile can also be fired. A total of 60 projectiles and charges are carried in carousel magazines each side of the breech area. Allied with a mechanical handling system this enables a fire rate of up to six rounds a minute.

152 mm 2S5

Specification
First prototype: early 1970s
First production: 1976
Current user: CIS
Crew: up to 7
Weight in action: 28200 kg
Length overall: 8.33 m
Width: 3.25 m
Height: 2.76 m
Road range: 500 km
Fording: 1.05 m
Powerpack: supercharged multi-fuel developing 520 hp
Depression/elevation: -2°/+57°
Traverse: 30°

152 mm 2S5 Self-propelled guns in action.

152 mm Self-propelled Gun 2S19 Former Soviet Union

The **152 mm 2S19 self-propelled gun**, sometimes referred to as the **MSTA-S**, was originally developed to be the replacement for the virtually unprotected 2S5 self-propelled gun (see previous entry), although the original plans appear to have been disrupted by the break-up of the old Soviet Union. Although production of the **2S19** continues at the Uraltransmash Works, much of the output is apparently destined to be diverted for export sales as the **2S19** is being actively marketed throughout the Middle East and elsewhere. Production commenced in 1989.

The **152 mm 2S19** is an alliance of the 40-calibre (approx) ordnance of the 152 mm 2A65 towed howitzer (qv) with a large turret mounted on a much-modified T-80 MBT chassis, with some automotive features taken from the T-72 MBT. The **2S19** therefore fires the same ammunition family as the towed 2A65 with a standard FRAG-HE projectile having a maximum range of 24000 metres, although a FRAG-HE with a base bleed (BB) unit added can improve this to 29000 metres. Other 152 mm ammunition innovations introduced for the **2S19** and 2A65 include a cargo round containing 42 dual purpose (anti-personnel and armour-penetrating) bomblets, each containing 45 grams of explosive. Another **2S19** projectile dispenses small radio transmitter bodies intended to jam enemy communications. The **2S19** can also fire the laser-guided Krasnopol anti-armour missile. Tactical nuclear and chemical projectiles have now been withdrawn.

The **2S19** turret is mounted centrally on its tank-derived chassis but the turret bustle protrudes to the rear sufficiently for a projectile transfer arm to take rounds from outside the turret and place them directly into the autoloader system. When this feature is not in use an automated ammunition selection and loading device can take and load projectiles from racks in the turret bustle, enabling a fire rate of up to eight rounds a minute to be achieved.

One unusual feature of the **2S19** is a rear-mounted snorkel device which, when fitted, enables the **2S19** to cross water obstacles up to 5 metres deep.

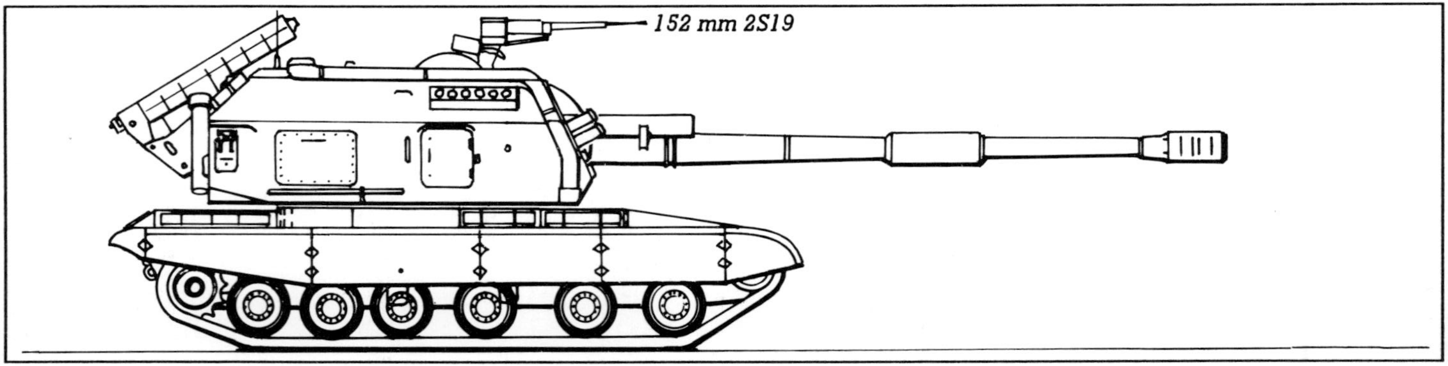

152 mm 2S19

Specification

First prototype: mid-late 1980s
First production: 1989
Current user: CIS
Crew: 5
Weight in action: 42000 kg
Length overall: 11.917 m
Length of hull: approx 7 m
Width: 3.38 m
Height: 2.985 m
Road range: 500 km
Fording: basic 1.5 m;
with preparation 5 m
Powerpack: V-84A diesel developing
780 hp
Depression/elevation: -3°/+68°
Traverse: 360°

152 mm Self-propelled gun 2S19 showing ammunition loading system at back of turret.

203 mm SO-203 Self-propelled Gun

Former Soviet Union

The **203 mm SO-203 self-propelled gun** is still something of a mystery to the West as production versions have never been publicly displayed and many details are still unknown. Sometimes known in the West as the 203 mm M-1975, the **SO-203** is code-named the **Pion (Peony)** or **2S7**. The first production examples were delivered in 1975 and are still operated by the heavy artillery regiments attached to most CIS armed forces fronts. At one time it was reported that numbers had been delivered to the former Czechoslovakia and Poland.

The **SO-203** utilises a very large special chassis with an armoured crew cab overhanging the front. More of the seven-man gun crew travel in a hull compartment behind the overhanging cab, with the engine compartment behind them. The long ordnance (the 203 mm 2A44, exact length undisclosed) is mounted at the rear of the chassis, with a massive recoil spade over the hull rear, to be lowered hydraulically before firing; extra stability is produced by lowering the hull rear using the variable suspension. The gun barrel is devoid of muzzle brake or fume extractor and is provided with a screw breech and a power loading device coupled to an ammunition handling system which enables a fire rate of two rounds a minute to be achieved for short periods.

Despite the numerous stowage boxes provided along the sides of the **SO-203's** hull, the chassis carries only four rounds ready for use. More ammunition is carried in trucks following the **SO-203**, (late production versions, known as the SO-203M, can carry as many as eight projectiles). The standard separate-loading HE round weighs 110 kg and has a maximum range of 37500 metres; a rocket-assisted (RAP) version weighs 102 kg and can reach 47000 metres. Other reported projectiles include concrete-piercing, chemical and tactical nuclear, the latter two now reported to be withdrawn.

203 mm SO-203

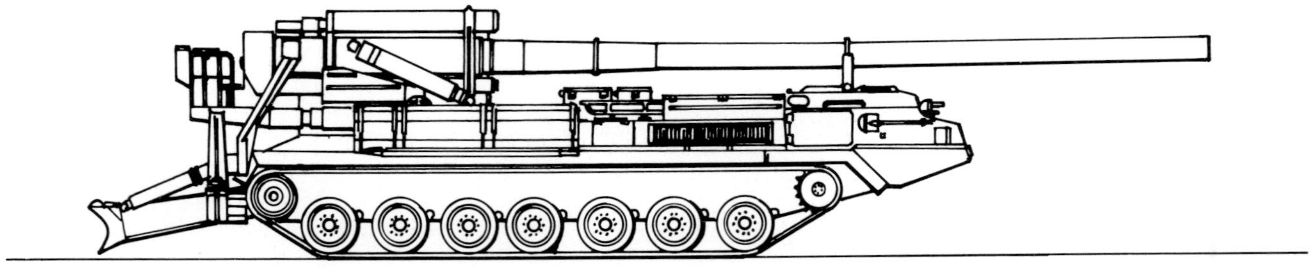

Specification
First prototype: early 1970s (?)
First production: 1974-1975
Current user: CIS
Crew: 7
Weight in action: 46500 kg
Length overall: 13.1 m
Width: 3.38 m
Height: 3 m
Road range: 675 m
Fording: 1.2 m
Powerpack: V-12 diesel developing 744 hp
Depression/elevation: 0°/+60°
Traverse: 30°

203 mm SO-203 Self-propelled Gun in travelling mode.

152 mm Self-propelled Gun-howitzer DANA

The **152 mm Self-propelled Gun-howitzer DANA** was developed during the late 1970s at a time when the Czech and Slovak Republics were one. Production started in 1981 using the Tatra 815 8 x 8 high mobility truck as the basis. Production of the basic chassis is still in progress.

Apart from the use of a wheeled chassis the layout of the **DANA** is unusual as the gun turret is located in the centre of the chassis behind a low slung armoured cab for the driver and commander. The rear of the chassis is occupied by the powerpack. Ammunition stowage is located around the hull. For firing, three stabiliser legs are lowered from the sides and rear.

The loading system is automatic. Projectiles are fed into one side of the split turret and charges into the other. From there they are moved to the breech area and power loaded. The barrel, which is about 39 calibres long, elevates between the two turret halves which is completely sealed against NBC agents.

The ammunition family used by **DANA** is similar to other modern Eastern Bloc 152 mm systems although many items are produced in the Czech and Slovak Republics. A standard HE projectile has a maximum range of 18700 metres - at least one type of bomblet-carrying cargo round is under development. About 60 projectiles and charges are carried by **DANA**.

A 47-calibre barrel variant, the ONDAVA, has been produced - a HE-BB projectile developed for ONDAVA has a maximum range of 32000 metres. Yet another variant is the 155 mm ZUZANA, produced in prototype form as a sales venture to fire NATO standard 155 mm ammunition. One further variant is an air defence vehicle armed with two 30 mm cannon and four surface-to-air missiles.

DANAs have been exported to Libya and Poland while a number are operated by some CIS states.

152 mm DANA

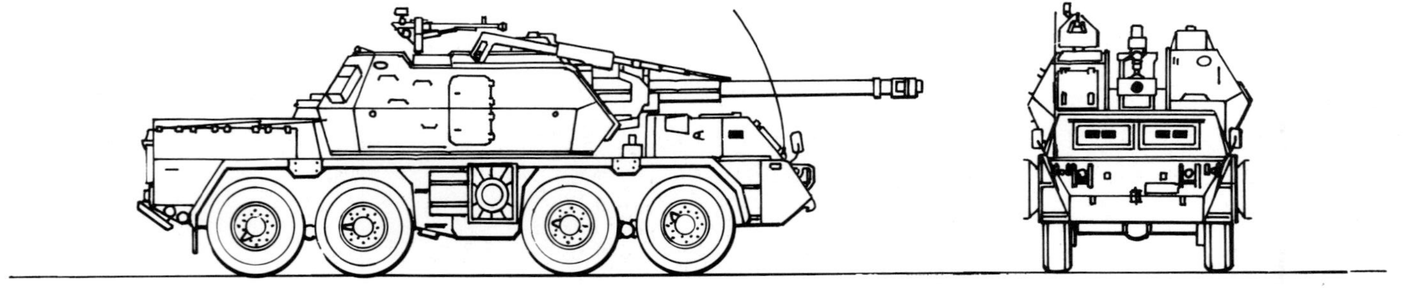

Specification
First prototype: late 1970s
First production: 1981
Current users: Czech & Slovak Republics, CIS, Libya, Poland
Crew: 5 or 6
Weight in action: 29250 kg
Length overall: 11.156 m
Width: 3 m
Height: 2.85 m
Road range: 650 km
Fording: 1.4 m
Powerpack: Tatra 2-939-34 diesel developing 345 hp
Depression/elevation: -4°/+70°
Traverse: 225°

152 mm Self-propelled Gun-howitzer DANA undergoing NBC decontamination.

155 mm Self-propelled Gun Mk F3

France

The **155 mm Self-propelled Gun Mk F3** is an example of a past generation of self-propelled artillery as it dates from the early 1950s, a time when crew protection was considered a relatively low priority. The **Mk F3** is essentially a 155 mm gun mounted directly onto a turretless AMX-13 light tank chassis with the chassis acting as a mobile carriage - only the driver has any protection so any crew members travelling with the gun have to perch on the top of the hull. Their normal means of transport is either in an accompanying truck or tracked carrier which also carries the ammunition.

Production of the **Mk F3** was carried out by Mecanique Creusot Loire, now part of Giat Industries; over 600 examples were manufactured. The 155 mm barrel is mounted at the extreme rear of the chassis with two trail spades manually lowered from the hull rear to provide extra stability when firing.

The 33-calibre **Mk F3** barrel fires a Mk 56 HE projectile with a maximum range of 20050 metres - the NATO standard HE M107 can reach 14700 metres. It is possible to fire enhanced range projectiles from the **Mk F3** barrel but this is not normally done due to limitations in the chamber capacity regarding the more powerful charges.

Updating conversion packages for the **Mk F3** have been introduced, one being the replacement of the original petrol powerpack by a diesel unit. A more drastic proposal includes the diesel unit but allied with a new 39-calibre barrel, the same as that on the towed 155 mm TR gun (qv). This version is the 155 AM 39 and would enable users to employ the latest enhanced range ammunition so maximum ranges could be 24000 to 31000 metres, depending on the type of ammunition involved.

The **Mk F3** was exported widely, one batch going to Argentina where local manufacture involved placing the barrel on a towed carriage to create the CITEFA Model 77, numbers of which were used in the Falklands conflict.

155 mm Mk F3

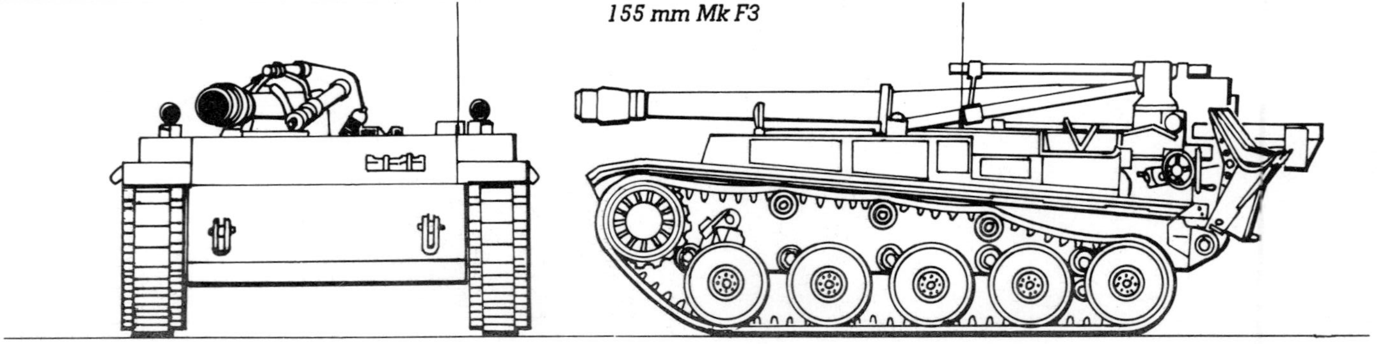

Specification

First prototype: early 1950s
First production: mid 1950s
Current users: France, Argentina, Chile, Cyprus, Ecuador, Morocco, Sudan, Qatar, UAE and Venezuela
Crew: 8
Weight in action: 17400 kg
Length overall: 6.22 m
Length of hull: 4.338 m
Width: 2.7 m
Height: 2.085 m
Road range: 300 km
Fording: 1 m
Powerpack: SOFAM Model 8Gxb petrol developing 250 hp
Depression/elevation: 0°/+50°
Traverse: 50°

155 mm Self-propelled Gun Mk F3 ready for action.

155 mm GCT Self-propelled Gun — France

The **155 mm GCT (Grand Cadence de Tir) self-propelled gun** was developed from 1969 onwards to replace all 105 and 155 mm self-propelled artillery then used by the French Army. The first prototype appeared in 1972, with production following in 1977 after a series of troop and other trials. Exports to Saudi Arabia commenced the following year. By the end of 1993 the French Army had taken delivery of over 250 **GCTs**.

The **GCT** is essentially a 155 mm 40-calibre gun turret on a modified AMX-30 tank chassis with the turret installed in place of the usual tank turret. There are only three of the four crew members in the turret (the other is the driver) as the **GCT** has an automatic loading system, enabling the gun to deliver six rounds in 45 seconds or eight rounds in one minute. A total of 42 projectiles and charges are stowed in racks in the turret bustle with re-supply taking place through large doors opening down from the rear of the turret. A 7.62 or 12.7 mm machine gun for local and air defence is mounted close to the commander's roof hatch.

A standard HE projectile fired by the **GCT** (the Mle 56/59) has a maximum range of 23000 metres. A similar projectile fitted with a base bleed (BB) unit can reach 29000 metres but the NATO HE M107 continues to be used, albeit with a maximum range of 18000 metres. Projectiles under development for the **GCT** include one dispensing anti-tank mines, a cargo round containing 63 dual purpose grenades, and a new family of HE, smoke and illuminating projectiles to replace those in service.

In theory the **GCT** turret could be installed on virtually any MBT chassis but to date other installations, apart from trials with a Leopard 1 MBT, have been limited to a demonstration model on a T-72 MBT chassis.

155 mm GCT

Specification

First prototype: 1972
First production: 1977
Current users: France, Saudi Arabia, Iraq, Kuwait
Crew: 4
Weight in action: 42000 kg
Length overall: 10.25 m
Length of hull: 6.7 m
Width: 3.15 m
Height: 3.25 m
Road range: 450 km
Fording: 2.1 m
Powerpack: Hispano-Suiza HS 110 multi-fuel developing 720 hp
Depression/elevation: -4°/+66°
Traverse: 360°

155 mm GCT Self-propelled Gun at maximum elevation during trials.

155 mm Panzerhaubitze 2000 (Pzh 2000) Germany

The **155 mm Panzerhaubitze 2000 (Pzh 2000)** is destined to be the main self-propelled artillery piece of the German Army of the future. It is still under development with production yet to commence. Even when in production the early anticipated requirements for 1254 units have been drastically reduced and totals for the Germany Army are not likely to exceed 238, with production extended over several years.

Two consortia produced prototype systems for the **Pzh 2000**, with the Wegmann and MaK entry being selected. A further four examples are to be produced before a production decision is made some time 'during the mid-1990s'.

The Wegmann/MaK **Pzh 2000** has its engine at the front and the welded steel gun turret mounted to the rear. The main ordnance is a 52-calibre piece with a multi-baffle muzzle brake and a fume extractor. An electrically-driven automatic loading system is used to transfer ammunition from a magazine holding 60 projectiles together with a 'unicharge' modular charge system stowed and loaded separately in equal segments - up to 288 segments are carried. Using the loading system burst fire rates of three rounds in 10 seconds are possible. Fresh ammunition is loaded into the vehicle via a hatch in the lower hull rear. Inside the turret are the commander and two loaders plus a gun layer although the fire control system involved can be virtually automated. On-board equipment includes a navigation and location system.

Crew protection includes a fire warning and extinguishing system, an NBC protection system, and the ability to fit active armour panels to the turret roof and sides.

The **Pzh 2000** can fire all NATO standard 155 mm ammunition, with a range of 30000 metres for standard projectiles and 40000 metres for enhanced range projectiles. Ammunition loads will include the new 'smart' projectiles currently in the development pipeline, examples being self-homing anti-armour munitions dispensed from cargo rounds, new cargo rounds with bomblets more powerful than those in service, and 'fire-and-forget' self-homing anti-armour projectiles.

155 mm PzH 2000

Specification

First prototype: 1989
First production: not yet in production
Current user: (planned) Germany
Crew: 4 or 5
Weight in action: 55000 kg
Length overall: 11.67 m
Length of hull: 7.87 m
Width: 3.48 m
Height: 3.4 m
Road range: 420 km
Powerpack: MTU 881 diesel developing 1000 hp
Depression/elevation: -2.5°/+65°
Traverse: 360°

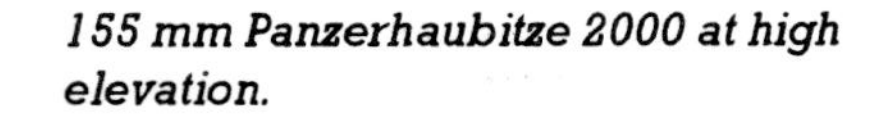

155 mm Panzerhaubitze 2000 at high elevation.

Palmaria 155 mm Self-propelled Howitzer — Italy

Most armoured vehicles are usually developed to meet a specific requirement but the **Palmaria 155 mm self-propelled howitzer** was developed from 1977 onwards by OTO Melara, specifically for possible export sales and to take full advantage of the then-current production run of OF-40 MBTs. The **Palmaria** subsequently has its 155 mm turret located centrally on a chassis derived from that of the OF-40. The ordnance design gained from Italian involvement in the International FH-70 towed howitzer programme (qv) so the **Palmaria** ordnance is matched ballistically with NATO standard 155 mm ammunition.

Export sales were not long in coming. In 1982 Libya ordered a batch of 251, with Nigeria also placing an order for 25, also in 1982. A batch of **Palmaria** turrets was ordered by Argentina to be involved in a locally-derived self-propelled artillery project known as the TAMSE VCA 155 but that project has yet to pass the prototype stage. Further **Palmarias** were produced in 1990 for an 'undisclosed customer'.

The chassis of the **Palmaria** is basically the same as that of the OF-40 MBT but a different engine is installed. The turret mounts the 39-calibre howitzer barrel which has both a fume extractor and a double-baffle muzzle brake. The turret has its own auxiliary power unit (APU) to power the turret service when the main engine is inactive. There is an automatic loading system with a total of 23 projectiles and charges located in the turret plus a further seven elsewhere around the hull.

Maximum rate of fire is three rounds in 20 seconds. A machine gun can be mounted on the turret roof.

The **Palmaria** can fire NATO standard 155 mm ammunition with the HE L15-type projectile range being 24700 metres; enhanced range base bleed (BB) projectiles can attain 30000 metres.

Proposed variants of the **Palmaria** include an autonomous 76 mm air defence gun turret and a twin 35 mm air defence gun system.

155 mm Palmaria

Specification

First prototype: 1977
First production: 1982
Current users: Libya, Nigeria
Crew: 5
Weight in action: 46000 kg
Length overall: 11.474 m
Length of hull: 7.265 m
Width: 3.35 m
Height: 2.874 m
Road range: 500 km
Fording: 1.2 m
Powerpack: MTU MB 837 Ea-500 developing 750 hp
Depression/elevation: -5°/+70°
Traverse: 360°

Palmaria 155 mm Self-propelled Howitzer ready for action.

155 mm Type 75 Self-propelled Howitzer — Japan

Ever since the establishment of the Japanese Self-Defence Forces it has been a policy to utilise locally-developed and produced defence materiel so, following the development of the 105 mm Type 74 self-propelled howitzer from 1969 onwards, consideration was directed towards a 155 mm equivalent. Two prototypes were manufactured in 1970 and 1971 with the final design standardised in 1975 as the **Type 75**. By the end of 1988 201 had been delivered and production ceased. Production of the **Type 75**, by Mitsubishi Heavy Industries, was expensive for many components and some of the assembly involved hand-fitting and finishing.

The **Type 75** uses a welded aluminium hull and turret with the main compartments in the 'classic' layout with a front-located engine and the turret to the rear. The howitzer involved is a Japanese design 30 calibres long and equipped with a double-baffle muzzle brake and fume extractor. Although the ordnance uses an interrupted screw thread breech there is a twin rotary magazine loading system with the barrel having to revert to an elevation angle of +6° for loading each round. Using this system the rate of fire is a possible six rounds a minute. The total ammunition load is the 18 rounds already in the loading system plus another ten projectiles and charges located around the turret interior. A 12.7 mm machine gun is located on the roof for local and air defence.

The Japanese have developed their own HE projectile for the **Type 75**, with a maximum range of 19000 metres. However the **Type 75** also fires the NATO HE M107 which reaches 15000 metres. A rocket-assisted projectile (RAP) with a planned range of around 24000 metres was under development at one time.

It is planned that the **Type 75** will be at least supplemented by a new 155 mm self-propelled howitzer mounted on a 39-calibre ordnance based on the FH-70 (qv), under licence in Japan

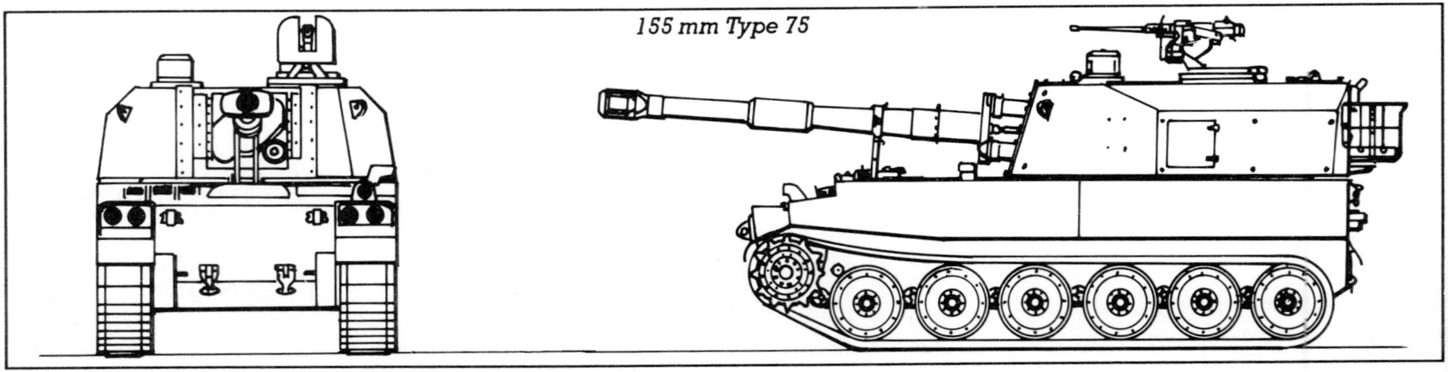

155 mm Type 75

Specification
First prototype: 1971
First production: 1975
Current user: Japan
Crew: 6
Weight in action: 25300 kg
Length overall: 7.79 m
Length of hull: 6.64 m
Width: 3.09 m
Height: 2.545 m
Road range: 300 km
Fording: 1.3 m
Powerpack: Mitsubishi 6ZF diesel developing 450 hp
Depression/elevation: -5°/+65°
Traverse: 360°

155 mm Type 75 Self-propelled Howitzer on manoeuvres.

G6 155 mm Self-propelled Howitzer — South Africa

When the South African Army selected the 155 mm G5 (qv) they accepted that the towed howitzer would not be able to keep up with their mechanised infantry columns. Accordingly a self-propelled version of the G5 was required and a wheeled carriage, selected to suit South African operational conditions, was developed to become the **G6 Rhino**. The first **G6** prototype appeared in 1981 with production beginning in 1988, only after pre-production examples had seen action in Angola.

The **G6** uses a large 6 x 6 chassis with the driver's compartment at the front and the 525 hp diesel behind him. The turret is at the rear, mounting the same 155 mm 45-calibre ordnance as the G5. Up to five crew are housed in the air-conditioned all-steel turret but normally one man stands outside to the rear, loading projectiles and charges into the auto-loader system one at a time through a small hatch in the hull rear. This allows the **G6** to maintain its full load of 45 projectiles and 50 charges for when rapid moves away from ammunition sources have to be made. Resupply of the loading system is through armoured slots in the hull rear. For firing, the chassis is raised onto four stabiliser legs, two at the sides and two at the rear, and practical traverse is limited to 40° each side.

The **G6** is part of an overall artillery system which includes the ammunition, based around the use of a wide array of ERFB projectiles which can be converted to ERFB-BB by fitting a base bleed (BB) unit in the field. Maximum ranges are 30000 metres for ERFB and over 39000 metres for ERFB-BB. The rest of the **G6** system includes a meteorology station, muzzle velocity analysers, a communications system and fire control, the latter being based around an all-electronic system which can be fully automated.

It is anticipated that 52-calibre barrels will be introduced to the **G6**.

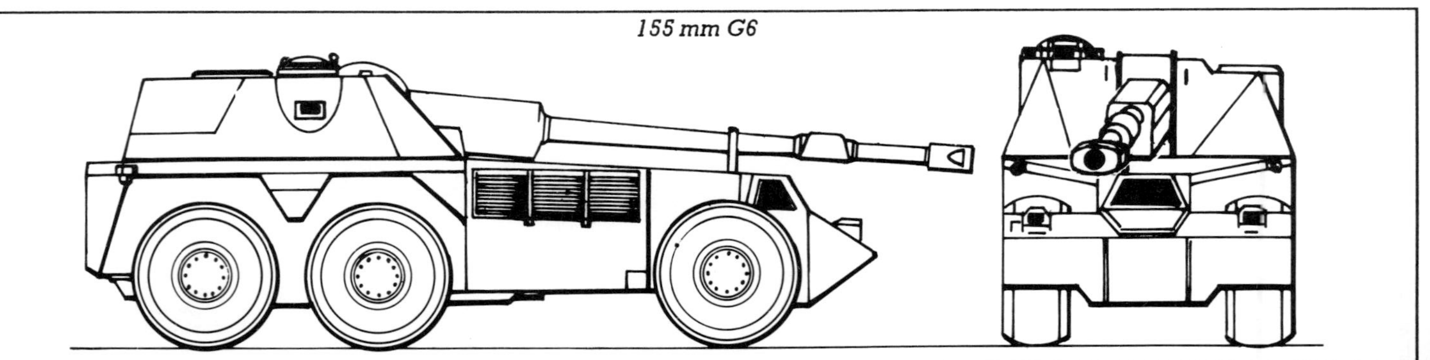
155 mm G6

Specification
First prototype: 1981
First production: 1988
Current users: South Africa, UAE
Crew: 6
Weight in action: 47000 kg
Length overall: 10.335 m
Length of hull: 9.2 m
Width: 3.4 m
Height: 3.3 m
Road range: 700 km
Fording: 1 m
Powerpack: diesel developing 525 hp
Depression/elevation: -5°/+75°
Traverse: 80°

G6 155 mm Self-propelled Howitzer with cupola accesses open.

155 mm Bandkanon 1A Self-propelled Gun

Sweden

The **155 mm Bandkanon 1A** was developed by Bofors during the late 1950s, with the first prototype appearing in 1960. In 1965 a production contract was awarded with the first example appearing the following year. Only 26 examples were produced for the Swedish Army but they are valued so highly that in April 1988 a further contract was awarded, again to Bofors, to modernise the guns and carriages, principally by the installation of new diesel engines coupled to automatic transmissions. Later improvements could include a new fire control and location system.

The **155 mm Bandkanon 1A** self-propelled carriage uses many components of the Bofors S-Tank series to the extent that the main automotive power is produced by a diesel engine and gas turbine with their outputs combined; the diesel is used all the time while the gas turbine is normally engaged only when moving across country. The rear of the chassis is occupied by the bifurcated gun turret, over which is the ammunition handling system. With this system a 14-round magazine over the rear of the turret and hull supplies fixed rounds to a loading tray which moves the rounds to the ramming position; the first round is loaded manually but all subsequent actions are automatic. With this system, which is completely protected by armour, 14 rounds can be fired within 48 seconds - the gunner can select either single shots or bursts. Once the main magazine is emptied it can be refilled through hatches in the top of the system mechanism using an integral hoist which lifts fresh rounds packed in clips; the re-loading operation takes about two minutes.

The gun is 50-calibres long and is fitted with a pepper-pot muzzle brake. Only one type of projectile is fired, the HE m/60 weighing 48 kg, with a maximum muzzle velocity of 865 m/s and a maximum range of 25600 metres.

155 mm Bandkanon 1A

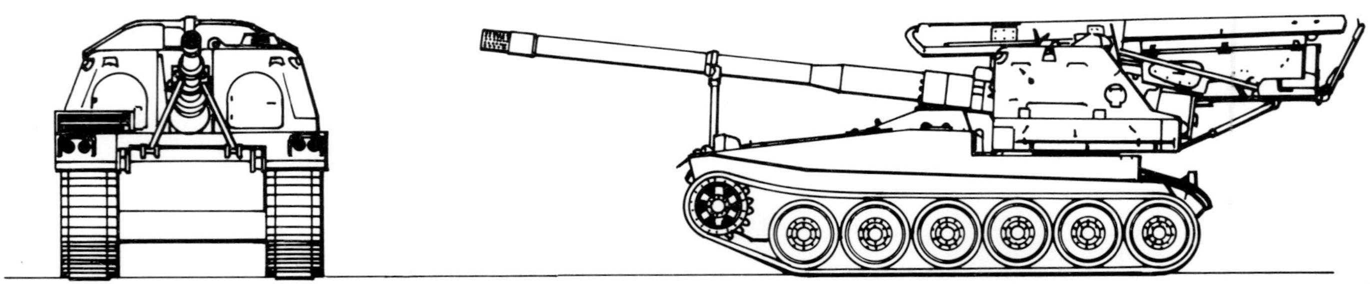

Specification
First prototype: 1960
First production: 1961
Current user: Sweden
Crew: 5
Weight in action: 53000 kg
Length overall: 11 m
Length of hull: 6.55 m
Width: 3.37 m
Height: 3.85 m
Road range: 230 km
Fording: 1 m
Powerpack: Detroit Diesel 6V-53T developing 290 hp plus Boeing gas turbine developing 300 hp
Depression/elevation: -3°/+40°
Traverse: (max) 30°

155 mm Bandkanon 1A Self-propelled Gun in travelling configuration.

155 mm Self-propelled gun AS90

UK

The **155 mm AS90** had a complicated development history as it was originally just a private venture turret with a 155 mm barrel matched ballistically with that of the FH-70 (qv) and intended for installation on virtually any MBT chassis to produce a relatively low cost self-propelled artillery system. The turret was the GBT 155, developed by Vickers Shipbuilding and Engineering Limited (VSEL) who later decided to produce a purpose-built chassis for their turret as a possible export venture. However the collapse of the international SP-70 155 mm self-propelled gun programme left the British Army with a requirement which was met, after further development, by the new GBT 155 and chassis combination known as **AS90**.

The first **AS90** prototype appeared in 1986, followed by others which had to undergo intensive trials to demonstrate they could meet the Army's requirements. Production commenced in 1991 with the first of 179 ordered being accepted for service in 1993. The **AS90** layout is of the 'classic' type with the powerpack and automatic transmission forward and the turret to the rear. The turret, which is modified from the GBT 155 original, has its own auxiliary power unit (APU) to provide power for the various turret services which includes electrical turret and barrel drives. On board systems include a fire control computer, a muzzle velocity analyser, and a navigation and location system, plus an NBC protection system. Ammunition is handled by an autoload system with which 31 pre-selected projectiles can be loaded from four magazine modules. More projectiles and charges are located around the turret and hull, making 48 in all.

The **AS90's** 39-calibre ordnance meets all NATO compatibility standards so can fire all NATO ammunition. Standard range is 24000 metres and 30000 metres with enhanced range base bleed (BB) projectiles.

A 52-calibre barrel may be retrofitted at some future stage along with increased automation leading to a reduction in crew strength

155 mm AS90

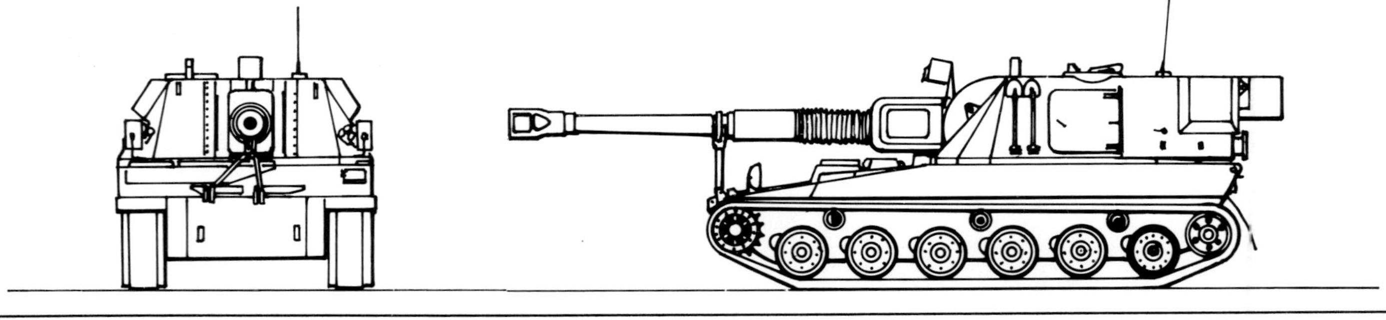

Specification

First prototype: 1986
First production: 1991
Current user: UK
Crew: 5
Weight in action: 45000 kg
Length overall: 9.9 m
Length of hull: 7.2 m
Width: 3.4 m
Height: 3 m
Road range: approx 370 km
Fording: 1.5 m
Powerpack: Cummins VTA 903T-660 V-8 diesel developing 660 hp
Depression/elevation: -5°/+70°
Traverse: 360°

AS90 155 mm Self-propelled Gun on Salisbury Plain.

155 mm M109 Self-propelled Howitzer Series USA

The **M109** series of 155 mm self-propelled howitzers is numerically the most important of all Western artillery systems having been in constant production since 1962 - the prototype appeared in 1959. Since then numerous **M109** marks, sub-marks and variants have been produced, not only in the USA but in many nations elsewhere.

The original **M109** shared the same chassis as the 105 mm M108 series (now virtually defunct). Using an aluminium armour hull, the basic **M109** has a turret at the rear with a 23-calibre 155 mm barrel. The spacious hull and turret are powered by reliable automotive components which are retained virtually the same to this day. Original **M109s** remain in service but most were later converted to **M109A1** or subsequent standards, mainly by the installation of a longer barrel.

From the **M109A1** to the **M109A5** standard most differences were slight. The long M185 barrel can fire the standard NATO HE M107 to 18100 metres compared to the original stubby barrel with its 14600 metres. Gradually the number of ammunition types grew and conversions or variants produced by other countries (Italy, Norway, Switzerland, Germany, etc) proliferated to a bewildering extent and are still being introduced.

The latest US Army version is the **M109A6 Paladin** produced, as with most of the rest of the **M109** series, by BMY. Some regard the **M109A6** as an exercise in squeezing the last drop of juice from the **M109** lemon as it is a 39-calibre piece ballistically matched with the M198 towed howitzer and thus other compatible NATO pieces and ammunition, including the latest powerful propellant charges. Numerous other detail changes have been introduced with the **M109A6** but the most important feature is that standard projectiles can be fired to a range of 30000 metres. One trials vehicle has a 52-calibre barrel installed.

Two of the many **M109** variants include an ammunition supply vehicle and a fire direction centre, both based on the **M109** chassis.

155 mm M109A1

Specification (M109A2)
First prototype: basic M109 1959
First production: basic M109 1962
Current user: well over 25 nations
Crew: 6
Weight in action: 24950 kg
Length overall: 9.12 kg
Length of hull: 6.19 m
Width: 3.15 m
Height: 3.28 m
Road range: 350 km
Fording: 1.07 m
Powerpack: Detroit Diesel 8V-71T diesel developing 405 hp
Depression/elevation: -3°/+75°
Traverse: 360°

British Army 155 mm 109A3 Self-propelled Howitzer during training operations.

203 mm M110 Self-propelled Howitzer Series USA

During the late 1950s the US Army initiated the development of a self-propelled artillery system using a common tracked chassis which could mount either a 155, 175 or 203 mm piece. By the early 1960s the 155 mm project had been dropped but the 175 mm piece became the M107. The M107 had a long service career with many nations but has now been largely withdrawn by most of them, the main remaining user being Israel.

The 203 mm project resulted in the **203 mm M110**. Originally this had a relatively short 203 mm barrel with an ancestry going back to before 1918 but still powerful and efficient enough to be adopted by the US Army and many others.

The 175 and 203 mm pieces shared the same open-topped tracked self-propelled carriage and the barrels were interchangeable. For both the barrels were mounted towards the rear, without protection of any kind and with limited on-carriage provision for carrying the crew or ammunition. A large recoil spade is located at the rear and an ammunition handling and loading system is provided.

By 1976 a new longer 203 mm barrel had been produced and retrofitted to many existing **M110** systems, resulting in the **M110A1**. Fitting a double baffle muzzle brake resulted in the **M110A2**, the main variant still in service.

For the US and British armies the **M110** series was usually deployed as a 'shoot and scoot' nuclear delivery system. With the withdrawal of this capability and the introduction of the Multiple Launch Rocket System (MLRS - qv) the **M110A2s** were nearly all withdrawn but the type soldiers on with many nations as a powerful long range artillery piece.

The **M110A2** fires a 92.53 kg HE projectile, the M106, with a muzzle velocity of 711 m/s and a maximum range of 22900 metres. Other projectile types were developed but remain little used.

The carriage of the M107/M110 was developed to become the M578 light armoured recovery vehicle.

203 mm M110

Specification (M110A2)
First prototype: 1978
First production: 1978
Current users: at least 15 countries
Crew: 5
Weight in action: 28350 kg
Length overall: 10.73 m
Length of hull: 5.72 m
Width: 3.149 m
Height: 3.143 m
Road range: 523 km
Fording: 1.066 m
Powerpack: Detroit Diesel 8V-71T diesel developing 405 hp
Depression/elevation: -2°/+65°
Traverse: 60°

US Army 203 mm M110 Self-propelled Howitzer with weather cover.

ASTROS II Multiple Rocket Systems

The Brazilian Avibras Industria Aerospacial S/A have produced several artillery rocket systems, the most successful of which has been the **ASTROS II** series (Artillery SaTuration ROcket System). Originally produced for the Brazilian Army in 1982, the **ASTROS** series has been exported to Iraq, Saudi Arabia and some Middle East nations.

The **ASTROS II** is a complete rocket system, involving not just the rockets and launchers but ammunition supply vehicles, a command and control vehicle, a fire control vehicle, and several types of maintenance and repair vehicle. The basis for the system is a Mercedes-Benz 6 x 6 10-tonne truck chassis with armoured cabs.

The launch vehicle carries its rockets in an armoured launch container with the rockets launched over the cab. Rockets are launched direct from their transport containers which are loaded through armoured hatches over the launch container. Fresh containers are loaded onto the launch vehicle by an ammunition supply vehicle using a hydraulic crane.

The **ASTROS II** system lends itself to firing different rocket calibres from the same basic launch vehicle. By varying the calibre of rockets within the common container it is possible to meet various fire missions as required. The smallest rocket involved is the 127 mm SS-30 with a range of from 9 to 30 km; a single launch container carries 32 of them, each weighing 68 kg. Then comes the 180 mm SS-40 weighing 152 kg and with a range of from 15 to 35 km; 16 are carried in a single launch container. Top of the range is the 300 mm SS-60 weighing 595 kg, ranging from 20 to 60 km; four are packed into a launch container.

Warheads, apart from HE, can contain an incendiary element, various land mines, or a number of delayed action bomblets to act as area denial weapons.

A typical **ASTROS II** battery has six launch vehicles, six resupply vehicles and a fire control vehicle. A command and control vehicle, together with maintenance vehicles, would be held at a higher command level.

A coastal defence rocket based on the **ASTROS II** has been proposed.

ASTROS II

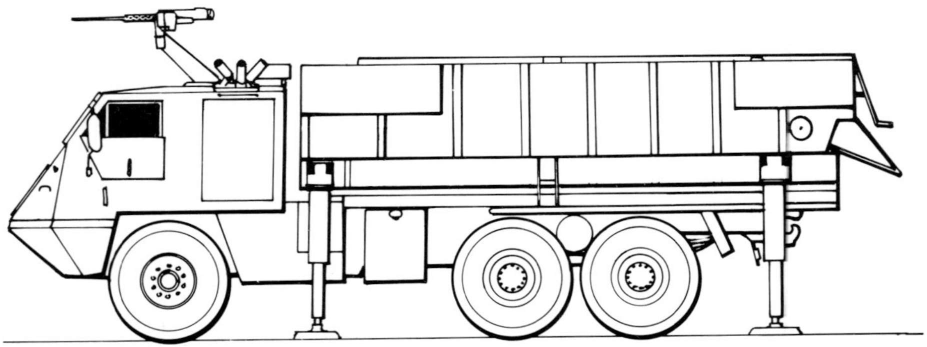

Specification

ASTROS II rockets			
Type	SS-30	SS-40	SS-60
Calibre (mm):	127	180	300
Number of tubes:	32	16	4
Rocket wt . (kg):	68	152	595
Rocket length (m):	3.9	4.2	5.6
Min range (km):	9	15	20
Max range (km):	30	35	60

Current users: Brazil, Iraq, Saudi Arabia

ASTROS II firing single rocket.

107 mm Type 63 Multiple Rocket System — China

Chinese **107 mm rocket systems** are among the most widely encountered of all artillery rocket systems as they have been liberally distributed, often to irregular forces such as the Afghan Mojahedin or Palestinian guerrillas. By far the largest users are the Chinese armed forces.

First produced during the late 1950s the most widely used launcher is the **Type 63**, a 12-barrel launcher having the tubes arranged in a box fashion with three layers of four tubes on a single axle split trail towed mounting. The **Type 63-1** is a lightened version of the **Type 63** while a man-pack version has been developed for use by airborne or mountain forces. Another launcher is known as the Type 81, basically a **Type 63** launcher carried on the rear of a light 4 x 4 truck with an enlarged crew cab to carry the four- or five-man crew plus 12 reload rockets. There is also the Type 85, a tripod-mounted single tube man-portable launcher, often distributed to guerrilla units as it weighs only 22.5 kg. Some irregular forces have launched 107 mm rockets after placing them directly on heaps of earth pointing in the required direction.

Using the **Type 63** launcher all 12 rockets can be fired within seven to nine seconds. Reloading takes about three minutes.

The 107 mm rocket is spin-stabilised and between 800 and 900 mm long, according to type; the usual weight is just over 18 kg. Types available include various types of HE, high fragmentation HE, HE incendiary and communication jammers. Range of a typical HE rocket is 8500 metres.

The Type 63 launcher has been produced outside China in North Korea, Iran and South Africa. The South African launchers were reverse engineered by Mechem Developments, not for local use but for export to various undisclosed nations. Known as the RO 107, the South African rockets have explosive warheads surrounded by steel balls set in epoxy resin to improve on-target effects.

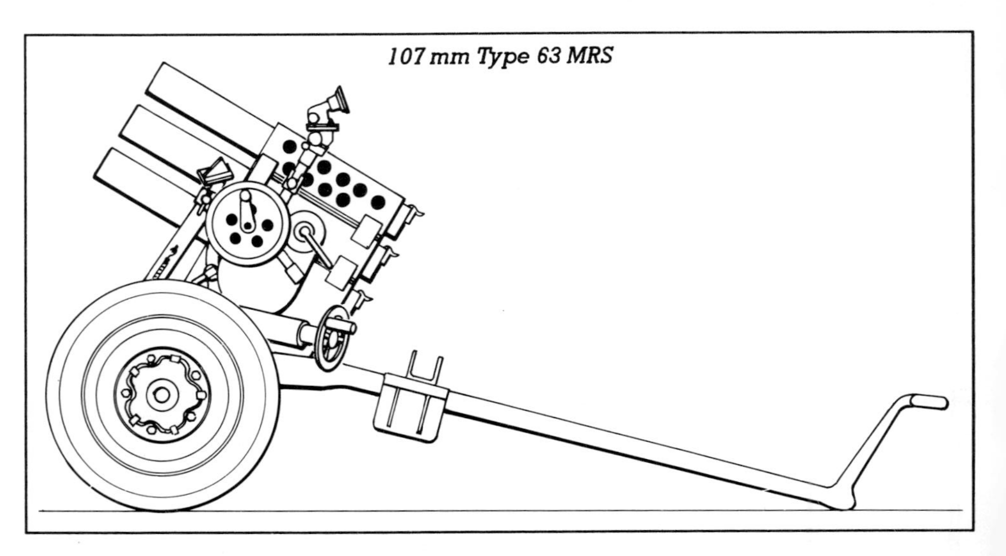

107 mm Type 63 MRS

Specification (Type 63)
First prototype: late 1950s
First production: 1959(?)
Current users: China, Iran, North Korea, Syria, Albania, Vietnam, Cambodia, Zaire, Chad and various guerilla forces
Crew: 5
Weight in action: 611 kg
Length travelling: 2.6 m
Width travelling: 1.4 m
Height travelling: 1.2 m
Max range: (typical) 8500 m
Depression/elevation: 0°/+60°
Traverse: 30°

Launcher for 107 mm Type 63 Multiple Rocket System ready for action.

122 mm Type 83 Multiple Rocket System — China

The **122 mm Type 83 multiple rocket system** is carried on the same hull and chassis as the 152 mm Type 83 self-propelled gun howitzer (qv). Towards the rear of the upper hull and in place of the turret the rocket-carrying **Type 83** has a roof with a turntable carrying 40, 122 mm rocket barrels arranged in four layers of ten tubes. Under the roof is a crew compartment for four of the crew - the other crew member is the driver seated at the front left of the hull.

The **Type 83** launcher is very similar to that used on the CIS 122 mm BM-21 (qv) and can be traversed and elevated using hydraulic power. All 40 rockets can be launched in one ripple salvo lasting 20 seconds although it is possible to fire each tube individually. After a full salvo has been fired it is then possible to reload the launcher by aligning the barrels with a reload pack carried over the front of the hull. Once the barrels are aligned with the pack a complete load of 40 rockets is automatically transferred to the barrels ready for another firing.

Several types of 122 mm rocket (actual calibre 122.4 mm) are available for use with the **Type 83** system, most of them with a maximum range of 20580 metres. The base warhead is HE with variations being HE-fragmentation, with the warhead surrounded by 4100 steel balls, HE-incendiary with 6000 small incendiary pellets scattered over a radius of 30 metres, and one further HE warhead with both fragmentation and incendiary effects. There is also a cargo rocket containing 39 dual-purpose bomblets which are scattered once a proximity fuze has functioned, and two types of mine-laying rocket which contain either anti-tank or anti-personnel mines. One of the latter, the Type 84 can carry either eight anti-tank or 128 anti-personnel mines to a range of 7000 metres.

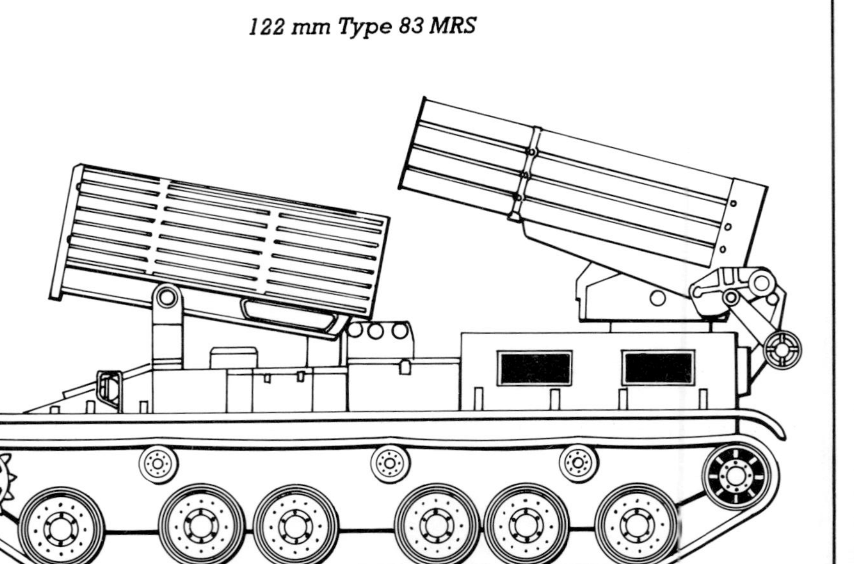

122 mm Type 83 MRS

Specification

First prototype: 1980(?)
First production: 1983
Current user: Ready for production
Crew: 5
Weight in action: 29900 kg
Length overall: 7.18 m
Length of hull: 6.85 m
Width: 3.145 m
Height: 3.18 m
Road range: 450 km
Fording: 1.3 m
Powerpack: Type 12150L V-12 diesel developing 520 hp
Max range: 20580 m
Depression/elevation: o°/+55°
Traverse: 168°

122 mm Type 83 Multiple Rocket System in rough terrain.

130 mm Type 82 Multiple Rocket System — China

The **130 mm Type 82** is produced by the China North Industries Corporation (NORINCO) and is a development of an earlier rocket system, the 130 mm Type 63. Both systems are based around the same 130 mm rocket (actual calibre 130.43 mm), the Type 63 HE which weighs 33 kg and has a range of 10100 metres. The difference between the Type 63 and **Type 82** launchers is that the Type 63 has 19 barrels arranged in two layers; the **Type 82** has 30 barrels arranged in three layers of ten. Type 63 launchers were mounted on 4 x 4 trucks or Type YW 531 tracked carriers. By contrast the **Type 82** is carried on a 6 x 6 truck or the Type 85 tracked carrier - the latter version is also known as the Type 85. For special or guerrilla forces a tripod-mounted single-barrel launcher has been produced which can be broken down into man-pack loads.

The 130 mm **Type 82** has the 30 barrels mounted over the rear two axles. The eight-man crew travel in an enlarged crew cab with 30 reload rockets in a locker located behind the cab. A full ripple salvo of 30 rockets can be fired in from 14.5 to 17.5 seconds with a second salvo being reloaded and fired within five minutes.

The Type 63 rocket, which is just over one metre long, has a high fragmentation warhead but a later type, the Type 82, has an enhanced fragmentation warhead created by 2600 small steel balls packed around the explosive. A further rocket, also known as the Type 82, contains 5000 incendiary pellets in place of the steel balls. There is one further rocket type used with the **Type 82** launcher, an extended range rocket which can reach 15000 metres. It has a high fragmentation warhead only.

130 mm Type 82 MRS

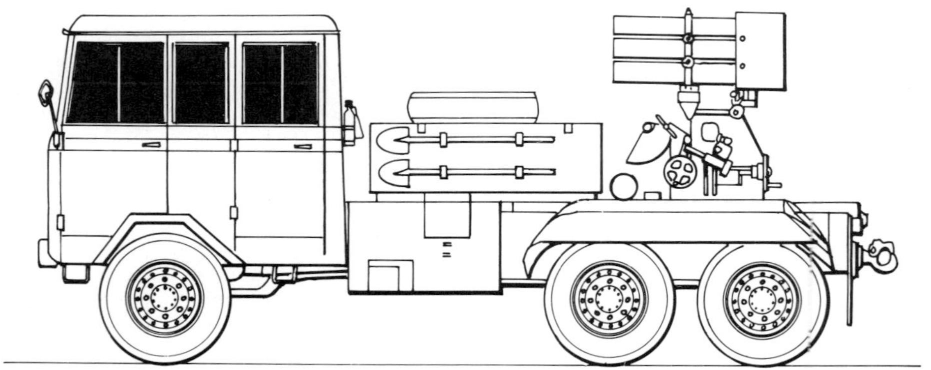

Specification (Type 82)
First prototype: early 1980s
First production: 1982
Current user: China
Crew: 8
Weight in action: 7500 kg
Length overall: 6.438 m
Width: 2.25 m
Height: 2.26 m
Maximum range: 10100 m
Depression/elevation: 0°/+50°
Traverse: 150°

130 mm Type 82 Multiple Rocket System in active configuration.

273 mm Type 83 Multiple Rocket System — China

The **273 mm Type 83 multiple rocket system** employs a fin-stabilised heavy rocket weighing 484 kg and with a maximum range of about 40000 metres. The **Type 83** system is carried on an unarmoured Type 60-1 tracked artillery tractor chassis with the launcher located in place of the usual load area. The system has four launchers arranged side-by-side. Before firing the launchers are manually elevated to +56°; a turntable provides a limited degree of on-carriage traverse. Also before firing two stabiliser legs are lowered from the rear of the chassis. Time to prepare for firing is about one minute. Re-loading the launcher after firing requires the assistance of a light crane or some other form of load handling device as it would be difficult for the five-man crew to handle the large rockets involved. All four rockets can be fired in 7.5 seconds.

The only known warhead type for the **Type 83** rocket is HE although it is very probable that other warheads such as chemical or some form of cargo payload could be involved. Each rocket is 4.753 metres long. The maximum velocity reached is 811 m/s providing a minimum range of 23000 metres.

Although it has been offered for export sales it is probable that the original **Type 83** is no longer in production as a development of the **Type 83** is used with the 273 mm WM-80 system. This system uses an eight-round launcher carried on an 8 x 8 heavy high mobility truck with the rockets having their range increased to over 80000 metres.

The rockets have an element of spin stabilisation as well as fins and have a warhead weighing 150 kg which can be either HE or cargo, the latter carrying 380 dual purpose bomblets. The WM-80 system includes a reload truck with a hydraulic handling crane.

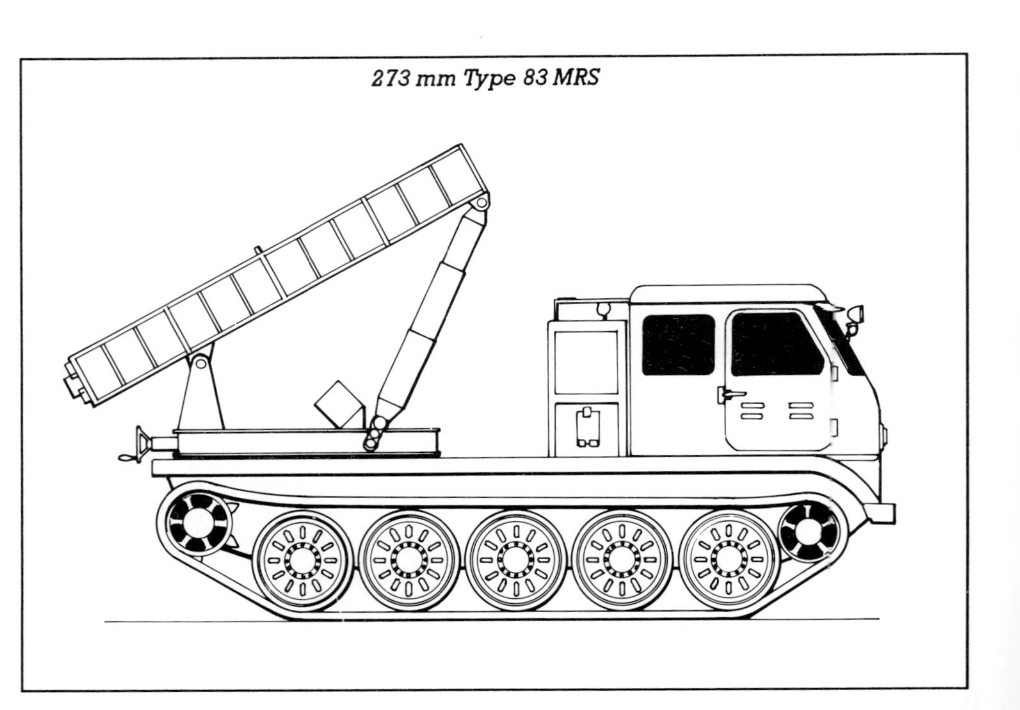

273 mm Type 83 MRS

Specification (Type 83)
First prototype: early 1980s
First production: 1983
Current user: China
Crew: 5
Weight in action: 17542 kg
Length travelling: 6.19 m
Width: 2.6 m
Height travelling: 3.18 m
Road range: 400 km
Fording: 1 m
Powerpack: Type 12150L V-12 diesel developing 300 hp
Max range: 40000 m
Depression/elevation: +5.5°/+56°
Traverse: 20°

273 mm Type 83 Multiple Rocket System at maximum elevation.

122 mm BM-21 Multiple Rocket System

Former Soviet Union

Development of the **122 mm BM-21** rocket system took place during the 1950s and since then it has become one of the most widely used artillery rocket systems in existence anywhere. It is produced in several forms, quite apart from the many clones which have been manufactured in several countries.

The base model is the **BM-21 Grad (Hail)** carried on Ural-375D 6 x 6 truck chassis; late production versions known as the **BM-21-1 Grad** (also known as the 9P137) use the diesel-engined Ural-4320 6 x 6 chassis. These carry 40 barrels. The Grad-V has only 12 barrels as it is mounted on a GAZ-66B 4 x 4 2-tonne truck chassis used by airborne forces. The Grad-1 uses the ZIL-131 6 x 6 truck chassis, carries 36 barrels and is also known as the 9P138. For special and guerrilla forces there is the single-barrel BM-21-P (9P132).

As a typical example the **BM-21 Grad** has 40 barrels mounted over the rear chassis of a Ural-375D truck with the barrels arranged in four horizontal layers of ten. Stabiliser legs are provided at the rear. All 40 rockets can be fired in 20 seconds.

Numerous types of **BM-21** rocket (actual calibre 122.4 mm) have been produced over the years. The latest family was introduced in 1991, having a maximum range of 30000 metres compared to the 20500 metres of earlier models, although rockets intended for the 36-round Grad-1 can reach only 15000 metres. Rockets for the single-barrel **BM-21-P** are shorter and can reach only about 11000 metres.

BM-21 warheads have included HE, HE-fragmentation, enhanced fragmentation, smoke, incendiary and chemical (now withdrawn). More recent types have included bomblet carriers and warheads containing anti-tank or anti-personnel mines.

A typical 122 mm rocket weighs up to about 77.5 kg and is between 2.8 and 3.25 metres long.

BM-21 series copies have been produced in China (Type 83 and Type 90), Egypt, India, Iran, Iraq, North Korea, Pakistan and Romania. Most of these nations also produce **BM-21** compatible rockets.

Italy's FIROS 25/30 (qv) is also **BM-21** compatible.

122 mm BM-21 MRS

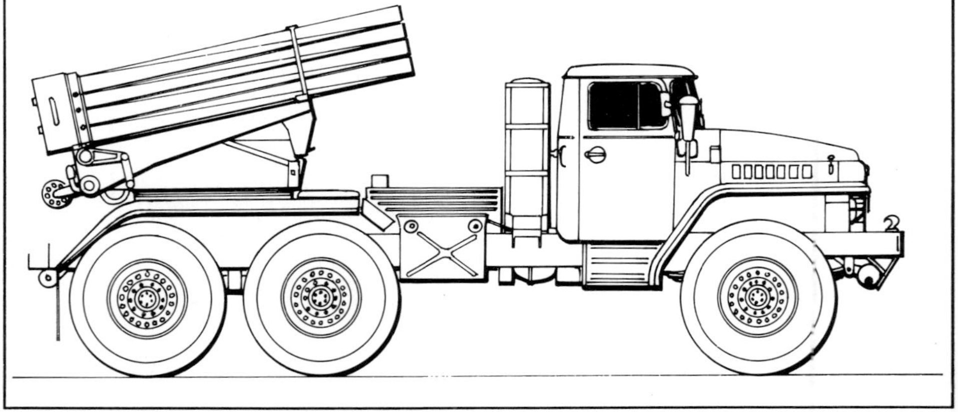

Specification (40-round BM-21 Grad)
First prototype: mid 1950s(?)
First production: 1962(?)
Current users: well over 30 nations
Crew: 6
Weight in action: 13700 kg
Length overall: 7.35 m

Width: 2.69 m
Height travelling: 2.85 m
Road range: 1000 km
Fording: 1.5 m
Powerpack: ZIL-375 V-8 petrol developing 180 hp

Max range: up to 30000 m
Depression/elevation: 0°/+55°
Traverse: (total) 180°

Battery of BM-21 Grads on the range.

140 mm BM-14-16 Multiple Rocket System

The **BM-14-16** rocket system dates back to the early 1950s and is no longer in front-line service with any of the CIS states, other than as reserves, although the type remains in large scale service with China and several other nations.

The **BM-14-16** has 16 barrels arranged in two horizontal layers of eight. The original carrier vehicle involved was the ZIL-151 6 x 6 truck chassis although later production versions use the more recent ZIL-131 6 x 6, 3.5-tonne chassis. Before firing can commence using either type of chassis armoured covers are lowered over the cab windows and two stabiliser legs are lowered at the rear. Four of the seven-man crew have to travel on the open area behind the cab.

The spin-stabilised 140 mm rockets (actual calibre 140.4 mm) fired by the **BM-14-16** are 1.092 metres long and weigh 39.62 kg. Maximum range is limited to 9810 metres and accuracy at the longer ranges is reported to be less than would be required from more modern equipments. Warhead types produced have included HE-fragmentation, smoke and chemical, although the latter is no longer used. The HE-fragmentation warhead weighs 18.8 kg.

Other 140 mm launchers using the same rockets as the **BM-14-16** include the BM-14-17 with 17 barrels carried on open-backed GAZ-63A 4 x 4 light truck chassis; these have largely been withdrawn other than by a few nations. Another possible launcher is the towed RPU-14 which has 16 barrels on a split trail launcher with four layers of four barrels; the RPU-14 was developed for issue to airborne units but they are now held in reserve by the CIS states. Another towed launcher was produced in Poland. This eight-barrel launcher is still issued to Polish airborne units and is known as the WP-8.

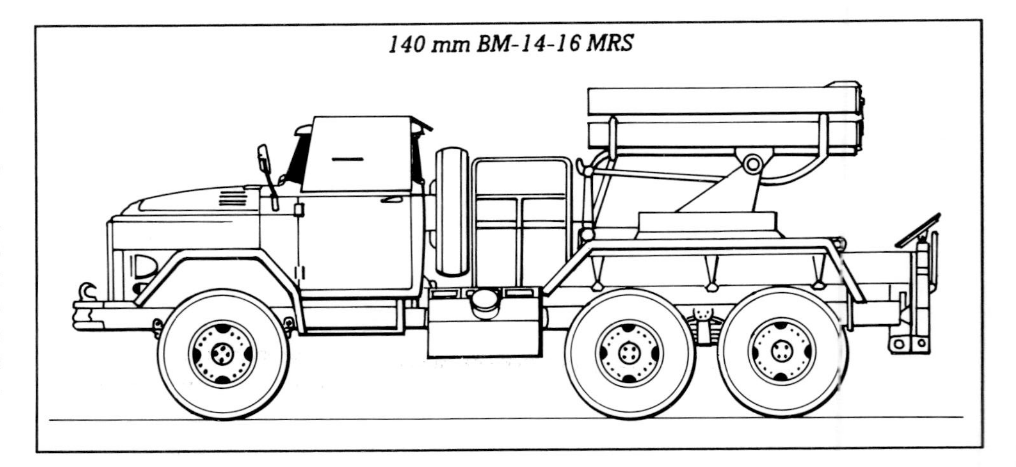

140 mm BM-14-16 MRS

Specification
First prototype: late 1940(?)
First production: early 1950s(?)
Current users: Algeria, Angola, Cambodia, China, Egypt, North Korea, Poland (WP-8), Somalia, Syria, Vietnam
Crew: 7
Weight in action: 8350 kg
Length overall: 6.92 m
Width: 2.3 m
Height travelling: 3.17 m
Road range: 600 km/h
Fording: 0.8 m
Powerpack: ZIL-121 petrol developing 92 hp
Max range: 9810 m
Depression/elevation: 0°/+52°
Traverse: (total) 200°

140 mm BM-14-16 Multiple Rocket System in travelling order.

220 mm BM 9P140 Uragan Multiple Rocket System

Former Soviet Union

For some years the **220 mm BM 9P140 Uragan (Hurricane)** multiple rocket system was something of a mystery to the West, being known by them as the Model 1977 or BM-22, but as the **Uragan** is now being offered for export sales all mystery has vanished.

The **9P140 Uragan** involves a 220 mm 16-round multiple launch rocket vehicle with one or two associated 9T452 reload vehicles, each incorporating a handling crane and a load of 16 rockets - the complete system is known as the 9K57. Both vehicles utilise the same modified 8 x 8 high mobility chassis of the ZIL-135LM twin-engined truck, with the launcher having the long barrels elevating from the extreme rear after two stabiliser legs have been lowered. The forward control cab of the **9P140** launcher vehicle houses most of the four-man crew, although some may travel on a reload vehicle, and all the firing preparations are made from within the cab.

The rockets fired by the **9P140 Uragan** have a maximum range of 35000 metres (minimum range 10000 metres). A typical 220 mm rocket, the HE 9M27F weighs 280 kg of which 100 kg is the warhead containing 51 kg of explosive. Each rocket is 5.178 metres long. Other rockets include a cargo warhead carrier containing 30 bomblets and two mine carriers, one for 24 anti-tank mines and the other carrying 312 of the infamous PFM-1 'Green Parrot' scatterable anti-personnel mines. It is known that chemical warhead rockets were produced for the Uragan.

A complete 16-round Uragan ripple salvo can be fired in 8.8 seconds, although 20 seconds is more normal, with a complete salvo covering an area of 426000 square metres. If required each rocket can be fired individually. After firing the reload time is from 20 to 30 minutes as each rocket has to be loaded and rammed individually.

220 mm BM 9P140 MRS

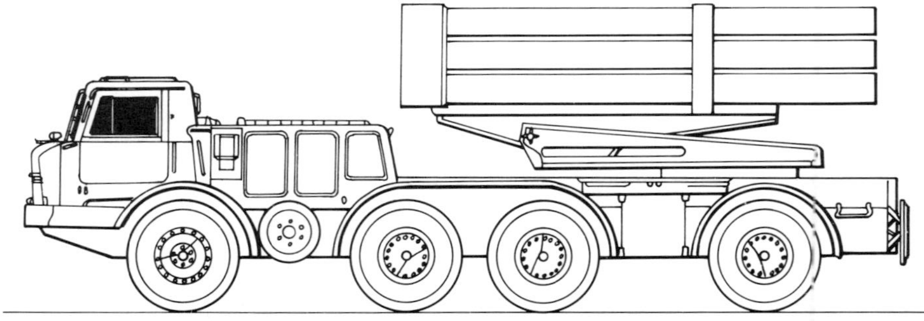

Specification (9P140)
First prototype: early 1970s
First production: mid 1970s(?)
Current users: CIS, Syria
Crew: 4
Weight in action: 20000 kg
Length in action: 10.83 m
Length travelling: 9.63 m
Width travelling: 2.8 m
Height travelling: 3.225 m
Road range: 500 km
Fording: 1.2 m
Powerpack: 2 x ZIL-375 V-8 diesels each developing 180 hp
Max range: 35000 m
Depression/elevation: +5°/+55°
Traverse: 60°

220 mm BM 9P140 Uragan Multiple Rocket System prepared for action.

300 mm BM 9A52-2 Smertch Multiple Rocket System

Former Soviet Union

The **300 mm BM 9A52-2 Smertch (Sandstorm)** multiple rocket system is the latest (export) version of the 9A52 Smerch (Tornado) system, at one time known in the West as the Model 1983. The **Smerch** was developed slightly after the Uragan (see previous entry) and is a much larger system overall, with a maximum range of 70000 metres (minimum range is 20000 metres). As with the Uragan the **9A52-2 Smertch** launch vehicle is part of an overall system known as the 9K58 which includes the **9A52-2 Smertch** launch vehicle plus at least one 9T234-2 reload vehicle. Also included in the system are 6 x 6 KamAZ-4310 battalion or battery command vehicles. The Smertch has been offered for export sales in the Middle East.

Both the 300 mm launch vehicle and the reload vehicle are based on 8 x 8 high mobility truck chassis, the MAZ-543A for the reload vehicle and the modified MAZ-543M in the case of the launch vehicle. The launch vehicle carries 12 lengthy launch tubes mounted at the rear of the chassis with stabiliser legs being hydraulically lowered between the third and fourth axles for firing. Firing a full 12-round ripple salvo takes up to 40 seconds, after which the launch vehicle retires to be reloaded by the 9T234-2 reload vehicle which carries a crane and a further 12 rockets. When fully loaded the reload vehicle weighs 30 tonnes.

The rocket involved with the **9A52-2** is the 300 mm 9M55K which contains a spin stabilisation element as well as fins to improve accuracy at the long ranges involved. The 9M55k has a bomblet-carrying warhead containing 72 HE-fragmentation sub-munitions - a complete 12-round salvo can cover an area of 672000 square metres with these bomblets. Each rocket is 7.6 metres long and weighs 800 kg. Operating the fire control equipment installed in the launch vehicle cab is eased for the four-man crew by the use of automated systems.

300 mm BM 9A52-2 MRS

Specification
First prototype: late 1970s
First production: early 1980s(?)
Current users: CIS
Crew: 4
Weight in action: 43700 kg
Length travelling: 12.1 m
Width: 3.05 m
Height travelling: 3.05 m
Road range: 850 km
Fording: 1.1 m
Powerpack: D12A-525 V-12 diesel developing 525 hp
Max range: 70000 m
Depression/elevation: +5°/+55°
Traverse: 60°

300 mm BM 9A 52-2 Smertch Multiple Rocket System jacked-up ready for action.

122 mm RM-70 Multiple Rocket System

Czech & Slovak Republics

The **122 mm RM-70** multiple rocket system was developed in the former Czechoslovakia during the late 1960s and was first observed by the West in 1972. It is essentially a close copy of the CIS 122 mm BM-21 launcher (qv) carried on the rear of a Tatra 813 8 x 8 high mobility truck - late production models use the Tatra 815.

Where the **RM-70** system differs from the BM-21 is its ability to utilise a rapid reload device positioned just behind the cab. With this device 40 rockets are held ready in open racks so that once a ripple salvo has been fired the empty launcher barrels are aligned with the racks for the fresh rockets to be pushed into the barrels ready for another firing. A complete reload sequence using the device takes about two minutes. The **RM-70** was the first launcher to use this facility which has since been copied for several other rocket systems, eg the former Yugoslav 128 mm 32-round 'Oganj'.

On early production versions of the **RM-70** the cab is armoured but with the introduction of the improved Tatra 815 chassis the cab was left unarmoured. These later versions are known as the Mod 70/85. Both the early and late production versions may be seen fitted with front-mounted hydraulically-operated dozer blades used to prepare firing positions or remove battlefield obstacles. Late production vehicles also have a winch and a central tyre pressure regulation system to improve cross-country mobility.

The fin-stabilised **122 mm RM-70** rockets closely resemble their CIS equivalents and are usually provided with HE-fragmentation warheads. Maximum range is about 20000 metres.

Variants of the **RM-70** system include a rocket-based minelaying system which can either have two 40-round rocket launchers or a single 40-round launcher and a mechanical system. The minelaying rockets can carry either anti-tank or anti-personnel mines. This system usually has an associated reload vehicle.

122 mm RM-70 MRS

Specification

First prototype: late 1960s
First production: 1970
Current user: Czech and Slovak Republics, Greece, Turkey, Poland, Libya, Zimbabwe,
Crew: 6
Weight in action: 33700 kg
Length overall: 8.8 m
Width: 2.5 m
Height travelling: 2.9 m
Road range: 600 km
Fording: 1.4 m
Powerpack: Tatra T-930-3 V-12 diesel developing 270 hp
Max range: approx 20000 m
Depression/elevation: 0°/+50°
Traverse: 172°

122 mm RM-70/85 Multiple Rocket System on Tatra 815 chassis.

110 mm Light Artillery Rocket System — Germany

The **110 mm Light Artillery Rocket System (LARS)** was developed during the 1960s with production commencing during 1970. Operated by only one country to date, Germany, some 209 systems were produced although they are now being phased from service as the Multiple Launch Rocket System (MLRS - qv) is introduced. When this happens it is anticipated that at least some **LARS** will be passed to Greece and Turkey.

LARS uses a 36-round launcher mounted on a MAN 6 x 6 7-tonne truck some of which are provided with armoured shutters over the windows. The launchers are arranged in two banks, each with 18 barrels, mounted on a turntable which can provide a full 360° traverse. Maximum range of the standard rockets is 14000 metres.

A full 36-rocket ripple salvo can be fired in 17.5 seconds. The usual procedure before a fire mission is to fire one rocket to be tracked by radar on a fire control vehicle with each battery. The radar is used to accurately determine the accuracy of an intended trajectory before a full salvo is released but this can be determined before the rocket is about half-way to the target so the rocket is then destroyed in flight before it can provide a warning to the target. Only after any necessary fire corrections are made is the full salvo launched. Late versions of **LARS** use an improved fire control system which can do away with the 'one rocket first' system. After a full firing the launch vehicles usually move to a new location as reloading takes about 15 minutes.

The **LARS** system involves a wide array of fin-stabilised rocket types. Warheads can include various types of screening smoke, high fragmentation with a point impact or proximity fuze, anti-tank mines with five AT-2 mines in each warhead, and a cargo warhead carrying bomblets (not yet in service). There are also two extended range rockets, one with a maximum range of 19000 metres, the other 25000 metres.

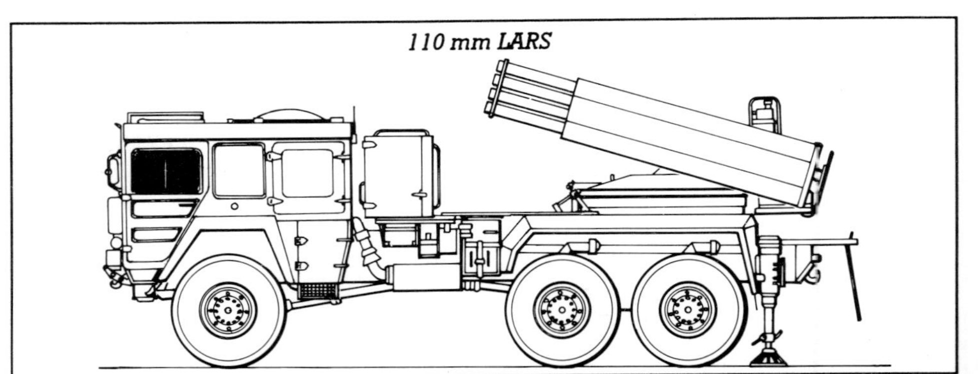
110 mm LARS

Specification
First prototype: mid 1960s
First production: 1970
Current user: Germany
Crew: 3
Weight in action: 17480 kg
Length overall: 8.28 m
Width: 2.5 m
Height travelling: 2.99 m
Road range: 500 km
Fording: 1.2 m
Powerpack: MAN diesel developing 260 hp
Max range: (standard) 14000 m
Depression/elevation: 0°/+50°
Traverse: 360°

110 mm Light Artillery Rocket System prepared for action.

227 mm Multiple Launch Rocket System

The **Multiple Launch Rocket System (MLRS)** was produced in response to a foreseen disparity in the artillery strength between NATO and the old Warsaw Pact. Originally a US Army programme, **MLRS** eventually became an international project involving the USA, UK, Germany, France and Italy (with participation from others), the prime US contractor being the Vought Corporation, now Loral Vought Systems.

MLRS launcher prototypes were manufactured during the late 1970s with the first US Army examples entering service in 1983. The first European-built launchers were delivered in late 1989. The launch vehicle is a tracked M270 carrier with an armoured cab for the three-man crew. Behind the cab is the Launcher Loader Module which carries two pre-loaded six-rocket pods loaded into the Module by a loader system on the vehicle. Once fired the pods are removed and discarded. If necessary all loading and other operations can be carried out from within the cab by one man.

The standard Phase 1 rocket weighs 307 kg and is 3.937 metres long; maximum range is about 32000 metres. The warhead contains 644 M77 bomblets which are released over a target to cover large areas with small shaped-charge high fragmentation munitions. A Phase 2 rocket contains 28 AT-2 anti-tank mines. Being lighter at 258.5 kg, Phase 2 rockets have a range of 40000 metres. Development of a Phase 3 rocket, which would have involved three sub-munitions with terminally-guided warheads, has been suspended, although the development of an extended range rocket, hopefully with a 45000-metre range, is continuing.

MLRS is one of the largest artillery projects currently under way in the West with orders for over 1000 launchers and 750000 rockets already placed.

MLRS developments include the Tactical Missile System (TACMS) for the US Army which can be launched from existing M270 launchers. This is a 610 mm rocket with a range of over 100000 metres - many of which were were used during the 1991 Gulf conflict.

227 mm Multiple Launch Rocket System

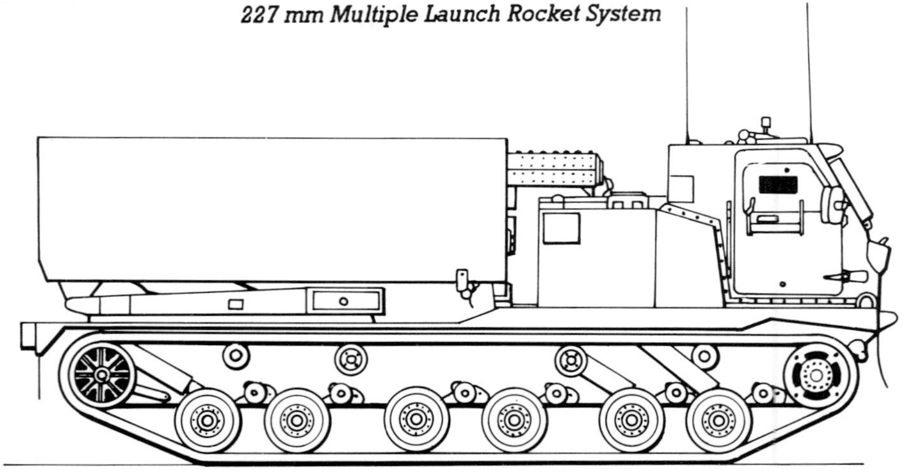

Specification
First prototypes: 1978
First production: 1982
Current user: USA, UK, France, Italy Germany, Netherlands, Bahrain, Japan, Saudi Arabia, Turkey.
Crew: 3
Weight in action: 25190 kg
Length overall: 6.972 m
Width: 2.972 m
Height travelling: 2.617 m
Road range: 483 km
Fording: 1.1 m
Powerpack: Cummins VTA-903 diesel developing 500 hp
Max range: (Phase 1) approx 32000 m
Depression/elevation: 0°/+50°
Traverse: 180°

Dutch army 227 mm Multiple Launch Rocket System in firing position.

160 mm LAR-160 Multiple Rocket System

Israel

The **LAR-160 artillery rocket system** first appeared during the early 1980s, produced by Israel Military Industries, now TAAS - Israel Industries. Designed to be a modular system, the **LAR-160** is based on the use of 160 mm fin-stabilised rockets packed into 18-round containers which are sealed at the factory and thereafter act as storage, transport and launch containers. On firing the rockets break their way through frangible end covers; after firing the containers are discarded.

The containers can be lifted in pairs onto a launcher carried either on a heavy truck, a suitable trailer or even a tracked chassis - **LAR-160s** sold to Venezuela were mounted on turretless AMX-13 light tank chassis although many other similar chassis could be utilised. Turretless M47 MBTs have been proposed as possible carriers, along with M548 tracked carriers. The original **LAR-160** rocket was the Mark 1, now replaced in production by the longer-range Mark 2. A fin-stabilised Mark 2 rocket is 3.314 metres long and weighs 110 kg - maximum range is 34000 metres compared to the 30000 metres of the Mark 1: the warhead weighs 46 kg; it was originally intended that warheads could be varied by attaching suitable and readily-available 155 mm artillery projectiles - hence the rocket diameter of 160 mm. However warheads produced to date are purpose-made and include the usual HE plus another containing 104 bomblet-type sub-munitions. One further type is known as a Pilot rocket which is fired to be tracked by radar prior to the launch of a full salvo. After the Pilot rocket has been tracked for part of its trajectory it is destroyed before it can warnthe target of an attack. Any necessary fire correction can then be made before a full salvo is released.

Development of the **LAR-160** system is continuing. One project involves the same containers but having two 350 mm rockets with a range of 80000 metres. Also planned are 160 mm warheads containing anti-tank mines or improved types of sub-munitions.

106 mm LAR-160 MRS

Specification (mounted on AMX-13)
First prototype: late 1970s
First production: 1980(?)
Current user: Israel, Venezuela
Crew: 5-6
Weight in action: 19200 kg
Length overall: 4.88 m
Length of hull: 4.88 m
Width: 2.51 m
Height: approx 2.5 m
Road range: 350-400 km
Fording: 0.6 m
Powerpack: Sofam Model 8Gxb petrol developing 250 hp
Max range: (Mark 2) 34000 m
Depression/elevation: 0°/+45°
Traverse: approx 170°

160 mm LAR-160 Multiple Rocket System being loaded.

FIROS 25/30 Multiple Rocket System

Italy

FIROS stands for FIeld ROcket System and it is produced in two versions, **FIROS 25** and **FIROS 30** the main difference between the two being the rockets. Both types of rockets have a diameter of 122 mm but their rocket motors differ, along with the maximum ranges possible.

The **FIROS** systems are produced by BPD Difesa e Spazio. There can be many differences in the type of system used by a particular customer, for apart from the rocket differences, the launch vehicles and their associated equipment can vary considerably. For instance a standard version can have either manually or partially motorised controls with all fire control equipment carried on a battery vehicle, while fully equipped systems can have refinements such as an on-board fire control computer, a navigation system and full power controls.

The launch vehicle is usually a 10-tonne 6 x 6 truck as it has to carry two launch modules. Each launch module carries 20, 122 mm rockets and is lifted onto the launch vehicle by a separate supply vehicle carrying up to four fresh modules and a handling crane.

FIROS rockets are fin-stabilised and, if required, can be fired from the CIS BM-21 system (qv). As a rough guide a **FIROS 25** rocket is 2.678 metres long and weighs 58 kg. A **FIROS 30** rocket is 2.815 metres long and weighs 65 kg. Maximum ranges are 25000 metres and 34000 metres respectively. (All with conventional warheads).

A wide variety of warhead is available for the **FIROS** rocket. Starting with conventional types there are controlled fragmentation or blast warheads, plus smoke for screening.

Various types of sub-munition warhead are available and are longer and heavier than the conventional warheads, usually having time or proximity fuzes to scatter their payloads over a target area while still in flight. Contents can include various types of bomblet or anti-tank mines.

FIROS 30 MRS

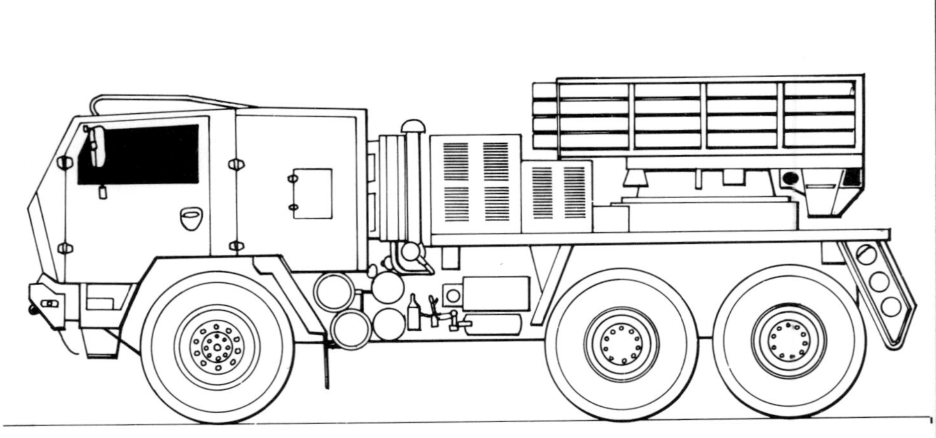

Specification

Crew: 2 or 3
Current users: Italy (FIROS 30), UAE (FIROS 25), plus others
Launch module weight:
F25 conventional 1566 kg;
F25 sub-munition 1806 kg
F30 conventional 1710 kg;
F30 sub-munition 1810 kg

FIROS 25/30 Multiple Rocket System ready to fire.

130 mm Type 75 Multiple Rocket System — Japan

In 1973, and following the usual national go-it-alone policy, the Japanese Self-Defence Technical Research and Development Institute produced two prototypes of an artillery rocket system which eventually became the **130 mm Type 75**. After a series of trials the **Type 75** was placed in production by the Aerospace Division of the Nissan Motor Company in 1975. Only 66 were produced for the Japanese Self-Defence Force. It is planned that the **Type 75** will eventually be replaced by the Multiple Launch Rocket System (MLRS - qv), 150 of which will be produced from kits in Japan, again by Nissan.

The **130 mm Type 75** uses a 30-round frame launcher in a rectangular container mounted over the hull of a much-modified Type 73 armoured carrier. Rockets may be fired either individually or in full 30-round ripple salvos, a complete salvo lasting 12 seconds. The launcher is controlled electrically but can be operated under manual control in an emergency. No reload rockets are carried so fresh rounds have to be carried in a following truck. A 12.7 mm machine gun is carried for air and local defence.

As far as can be determined only one type of rocket is launched from the **Type 75**, namely HE with each rocket being 1.856 metres long and weighing approximately 43 kg, of which 15 kg is the warhead; maximum range is 15000 metres. **Type 75** rockets are fin-stabilised and have an actual calibre of 131.5 mm.

To ensure maximum possible accuracy each **Type 75** battery is provided with a **Type 75** ground-wind measuring unit vehicle with an instrument-carrying mast to provide data for fire control. At battalion level there is the Type 76 artillery location radar vehicle to detect suitable targets. Both these ancillary vehicles make use of the same Type 73 tracked carrier chassis.

130 mm Type 75 MRS

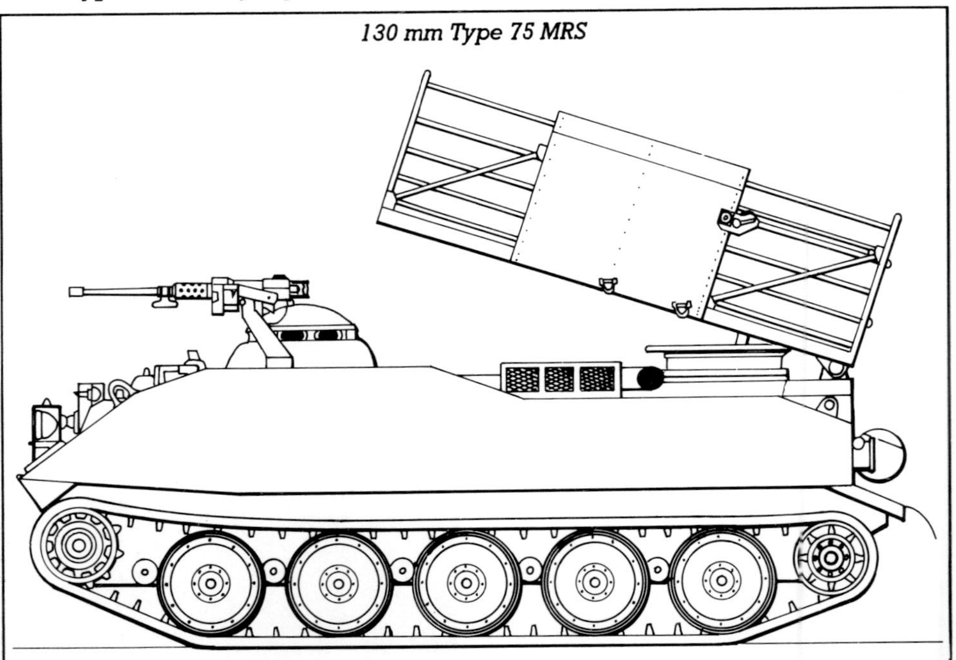

Specification

First prototype: 1973
First production: 1975
Current user: Japan
Crew: 3
Weight in action: 16500 kg
Length overall: 5.78 m
Width: 2.8 m
Height travelling: 2.67 m
Road range: 300 km
Powerpack: Mitsubishi 4ZF V-4 diesel developing 300 hp
Max range: 15000 m
Depression/elevation: 0°/+50°
Traverse: 100°

130 mm Type 75 MRS in action firing single rocket.

127 mm Valkiri Mark I 22 Multiple Rocket System

South Africa

During the late 1970s South Africa was in conflict with most of the border states to the north, many of whom were equipped with 122 mm BM-21 rocket systems (qv) from the Eastern Bloc. To counter the BM-21s, from 1977 Armscor developed their own 127 mm artillery rocket system which emerged in 1981 as the **Valkiri Mark I 22**. First deliveries to the South African Army began in 1982 and they went into action in Angola soon after.

The **Valkiri** uses a 24-round launcher carried on 4 x 4 Unimog trucks in such a manner that when the vehicles are on the move they resemble the ordinary cargo trucks used to carry reload rockets. Only when the covers are removed and the launchers elevated and/or traversed can their true nature be detected. On the launchers the 24 launch tubes, each 3 metres long, are arranged in three layers of eight. All 24 rockets can be fired within 24 seconds; re-loading takes 10 minutes.

The rockets fired are 2.68 metres long and weigh 53 kg. Maximum range is 22000 metres and minimum range 8000 metres - for the lower ranges spoiler rings are fitted to the rocket nose to reduce velocity. The warhead is high explosive with 8500 steel balls packed in resin around the explosive.

Each **Valkiri Mark I 22** battery has a command vehicle and a wind-measuring vehicle equipped with meteorological instruments.

In 1988 a short range towed version of the **Valkiri** was unveiled. Known as the Valkiri Mark I 5 the launcher involved has 12 tubes on a single-axle trailer towed by a light truck. For this system new rockets were developed with a maximum range of 5500 metres, each rocket being 1.4 metres long and weighing 30 kg. There is also a 32 kg rocket with a launch weight of 32 kg and a maximum range of 7000 metres. If required the Valkiri Mark I 5 can be used for direct fire against point targets.

Production of the Valkiri Mark I systems has ceased in favour of the later Mark II (see following entry).

127 mm Valkiri Mk I MRS

Model of 127 mm Valkiri Mark I 22 Multiple Rocket System prepared for maintenance check.

Specification (Valkiri Mark I 22)

First prototype: 1981
First production: 1982
Current user: South Africa
Crew: 2
Weight in action: 6400 kg
Length overall: 5.35 m
Width: 2.3 m
Height travelling: 2.32 m
Road range: 400 km
Fording: 0.6 m
Powerpack: diesel developing 99 hp
Max range: 22000 m
Depression/elevation: 0°/+50°
Traverse: 110°

127 mm Valkiri Mark II Multiple Rocket System

South Africa

Development of the **127 mm Valkiri Mark II** multiple rocket system to replace the Mark I systems (see previous entry) began in 1985 with production commencing in 1989. In time it is anticipated that the Mark II will replace the Mark Is in service.

The Valkiri Mark II is known to the South African Army as the Bataleur (Battler) and is mounted on a strengthened SAMIL 100 6 x 6 10-tonne truck chassis carrying 40 launch tubes arranged in two blocks of 20 on a turntable over the two rear axles. The cab for the five-man crew is armoured and has a V-shaped profile underneath to provide added protection against land mines. The cab and stowage lockers around the vehicle contain sufficient supplies, water, etc, to allow the vehicle to remain in the field for up to 14 days; extra fuel tanks for a 1000 km range are installed. A fully computerised fire control system is located in the cab, with peripherals such as navigation and location systems under development to ensure each vehicle can operate as an autonomous unit if required.

The **127 mm Valkiri II** rockets are 2.95 mm and weigh 62 kg. Maximum range is a nominal 36000 metres which can be reduced to 8000 metres if spoiler rings are fitted around the nose. Various warheads can be involved such as HE-fragmentation, anti-armour sub-munitions, and various types of bomblet. Fuzes involved can include point impact or proximity, including pre-set altimeter fuzes for the cargo warheads. Existing Valkiri Mark I rockets can be fired if required (see previous entry) although range is reduced.

A full load of 40 rockets can be fired in a ripple lasting 46 seconds, with stabiliser legs lowered at the chassis rear. Reloading can be undertaken using a raised platform at the rear.

If required the **Valkiri Mark II** can be mounted on vehicles other than the SAMIL 100.

127 mm Valkiri Mk II

Specification

First prototype: 1986(?)
First production: 1989
Current user: South Africa
Crew: 5
Weight in action: 21500 kg
Length travelling: 9.3 m
Width: 2.35 m
Height travelling: 3.4 m
Road range: 1000 km
Fording: 1.2 m
Powerpack: V-10 diesel developing 315 hp
Max range: 36000 m
Depression/elevation: 0°/+50°
Traverse: 110°

127 mm Valkiri Mk II Multiple Rocket System in action.

140 mm Teruel Multiple Rocket System — Spain

The **140 mm Teruel** artillery rocket system was developed by the Spanish Council for Rocket Research and Development with marketing and manufacture carried out by SANTA BARBARA. Although exact dates are uncertain it appears that development commenced during the mid-1970s with production starting during the early 1980s. The **Teruel** is in service with the Spanish Army and Gabon.

The **Teruel** has a 40-round launcher mounted on a modified Pegaso 3055 6 x 6 truck. An armoured crew cab for the five- or six-man crew is provided, with a 7.62 mm machine gun on the roof for local and air defence. The launcher tubes are positioned over the rear axles and arranged in two 20-tube blocks located inside two containers. Four hydraulic stabiliser legs are lowered, two each side, to raise the vehicle off its suspension for firing.

Two types of 140 mm spin-stabilised rocket (actual calibre 140.5 mm) are available, long and short. The long rocket has a range of 28000 metres, is 3.23 metres long and weighs 76 kg. Figures for the short rocket with its reduced rocket motor combustion time are 18000 metres, 2.044 metres and 56 kg. In both cases the warhead weighs 20 kg and can contain either HE-fragmentation, dual purpose (anti-personnel/anti-armour) or smoke sub-munitions, or six anti-tank mines. Spoiler rings can be fitted around the rocket nose to reduce range. Fuzes may be either point impact, electronic time or proximity.

Each **Teruel** battery has a number of supply trucks, each carrying packs of either 80 long rockets or 120 short. Reloading is carried out using a crane on the supply vehicle and usually takes about five minutes. A full 40-round salvo can be ripple-fired in 45 seconds. The time for a complete arrival at a firing site, preparation, firing and leaving is usually about five minutes.

140 mm Teruel MRS

Specification

First prototype: 1970s(?)
First production: early 1980s
Current user: Spain, Gabon
Crew: 5-6
Weight in action: approx 10000 kg
Length overall: approx 9 m
Width: 2.5 m
Height: approx 3 m
Road range: 550 km
Fording: 1.1 m
Powerpack: diesel developing 220 hp
Max range: long rocket 28000 m;
short rocket 18000 m
Depression/elevation: 0°/+55°
Traverse: 240°

140 mm Teruel Multiple Rocket System ready to fire.

70 mm Rapid Deployment Multiple Rocket Weapon System

USA

The **70 mm Rapid Deployment Multiple Rocket Weapon System**, or **RD-MRWS**, was a joint development undertaken during the 1980s by the US Army and BEI Defense Systems of Euless, Texas, to provide rapid deployment forces with a lightweight land-based ground support rocket system. Overall the intention was to deploy in-service 70 mm HYDRA 70 aircraft launch pods installed on ground launchers. Typical launch platforms proposed include converted 105 mm howitzer carriages, obsolete chemical rocket launch trailers, converted quadruple machine gun mountings, or purpose-built pedestal installations to be carried on light vehicles. Also developed was a tripod-mounted man-portable launcher for up to four rockets.

The basis of the **RD-MRWS** is a reloadable seven-round or 19-round launch pod originally intended for use with aircraft or helicopters. Up to eight of these pods can be mounted on a single towed mounting or vehicle installation with a control unit selecting how many rockets and/or pods could be fired at one time. A typical installation would be a six-pod launcher on a towed

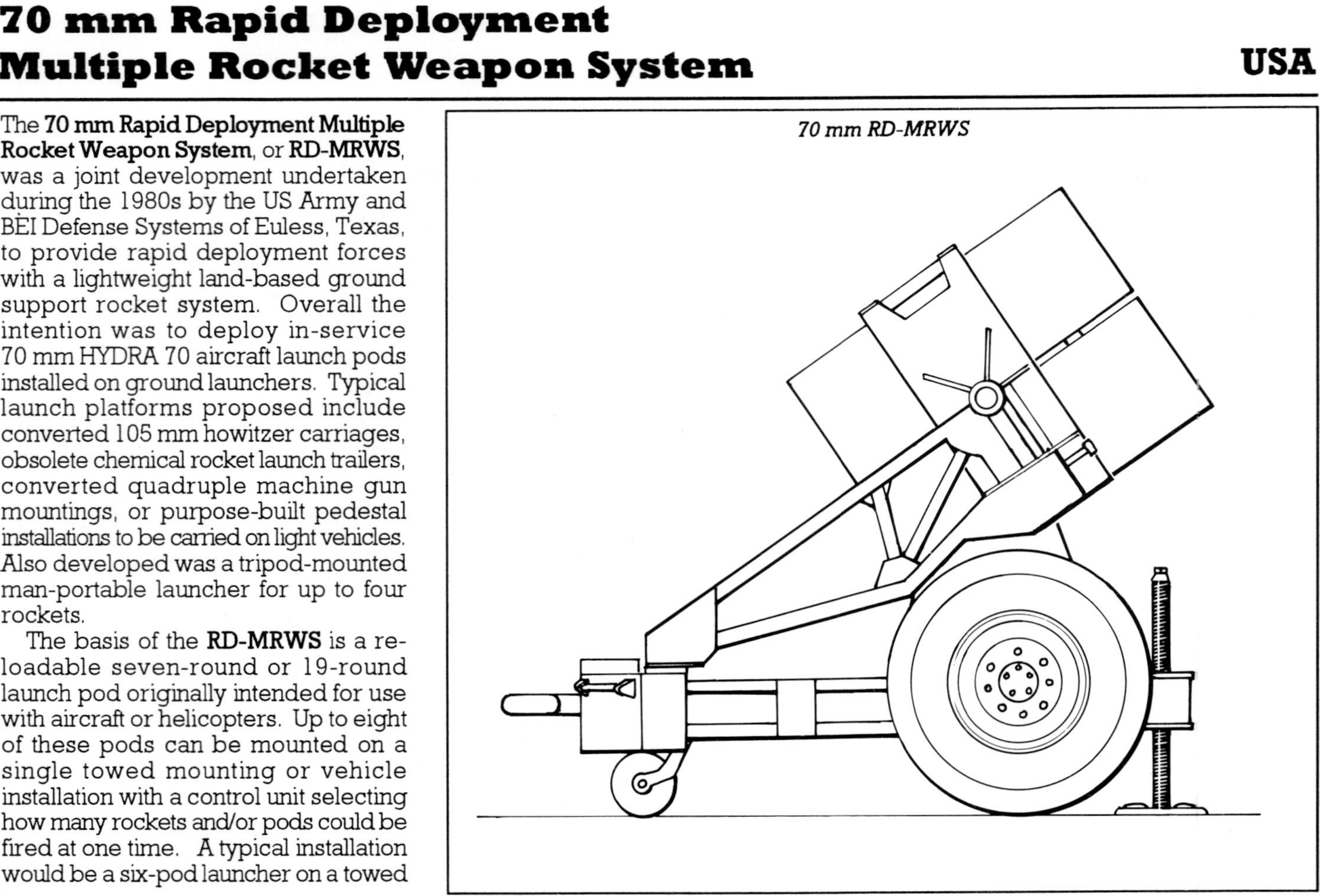

70 mm RD-MRWS

trailer, with a four-man crew taking about 10 minutes to load the pods. Rockets could then be fired either individually or in two or four-round salvos - if required all six pods could be emptied in less than seven seconds.

The choice of HYDRA 70 rocket types is wide. Starting with a standard HE warhead weighing 4.5 kg, other rocket types available include a cargo warhead carrying various types of multi-purpose sub-munition (some intended for use against helicopters), an anti-armour shaped charge warhead, at least two types of anti-personnel warhead containing flechettes, screening smoke, illuminating, and a number of radar-jamming chaff dispensers. Fuze options vary as widely as the warheads. Ranges vary from 700 metres to over 15000 metres.

The **RD-MRWS** was Type Classified by the US Army but no production has yet taken place. Instead the search for some other form of light artillery rocket system continues, the latest project involves a variant of MLRS (qv) with a single six-round 227 mm MLRS rocket pod carried on a 5-tonne truck.

Typical example of RD-MRWS.

Specification

Current user: Type classified by the US Army but not yet in production

82 mm Automatic Mortar 2B9 Vasilek

Former Soviet Union

Although the 82 mm mortar calibre is usually associated with infantry mortars it falls into the artillery category when the **82 mm 2B9 Vasilek (Cornflower)** automatic mortar is considered. This is due mainly to the practical fire rate of the **2B9**, which can be as high as 100 rounds a minute (120 rounds a minute cyclic), and its ability to fire at low elevation angles in a light support weapon or anti-armour role.

The **2B9** is normally mounted on a light wheeled carriage resembling a light field artillery mounting, complete with trails and a firing platform under the forward carriage. The centre of the barrel is mounted in a traversing drum-like assembly which acts as the upper carriage, through which a magazine-type housing enables standard 82 mm mortar bombs to be loaded in four-bomb clips from the right-hand side. The **2B9** can be fired as a mortar, using bombs loaded manually from the muzzle, at high barrel elevation angles, or for automatic direct fire in the manner of a light or anti-tank gun (direct fire sights are provided) using the breech magazine loading feature. To add to the complexity of the system, muzzle loading involves three possible variable propellant charges while breech loading involves a single fixed charge. Recoil forces are absorbed by a hydrospring recoil system around the barrel.

The **82 mm Vasilek** is also used on mobile mountings, one of the most common being carried on the load area of a GAZ-66 4 x 4, 2-tonne truck. Another self-propelled system is a Vasilek mounted inside the load area of a suitably modified MT-LB tracked carrier Hungary produces a similar self-propelled arrangement using a custom-modified MT-LB carrier.

The **2B9 Vasilek** fires standard 82 mm mortar ammunition with each bomb weighing about 3.1 kg - a special shaped charge warhead bomb has been developed for anti-armour use in the direct fire role. Maximum range as a mortar is 4200 metres.

It has been proposed that an 81 mm version of the **2B9 Vasilek** could be produced.

82 mm AM-2B9

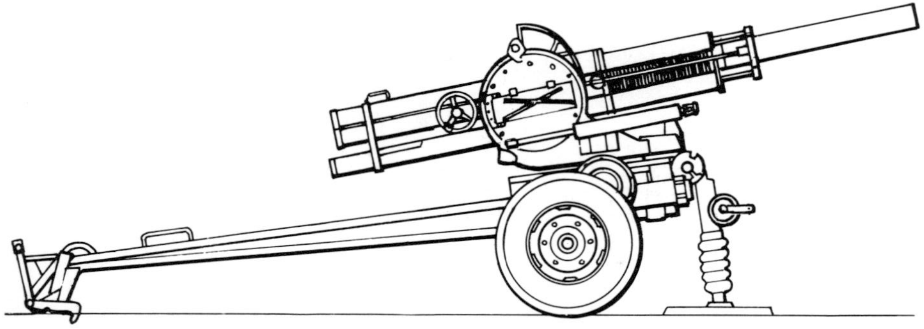

Specification
First prototype: late 1960s
First production: 1971
Current users: CIS, Hungary and others
Crew: 4

Weight in action: 632 kg
Max range: 4700 m
Depression/elevation: -1°/+85°
Traverse: 60°

82 mm Automatic Mortar 2B9 Vasilek. (Photograph from a Hungarian arms sales brochure).

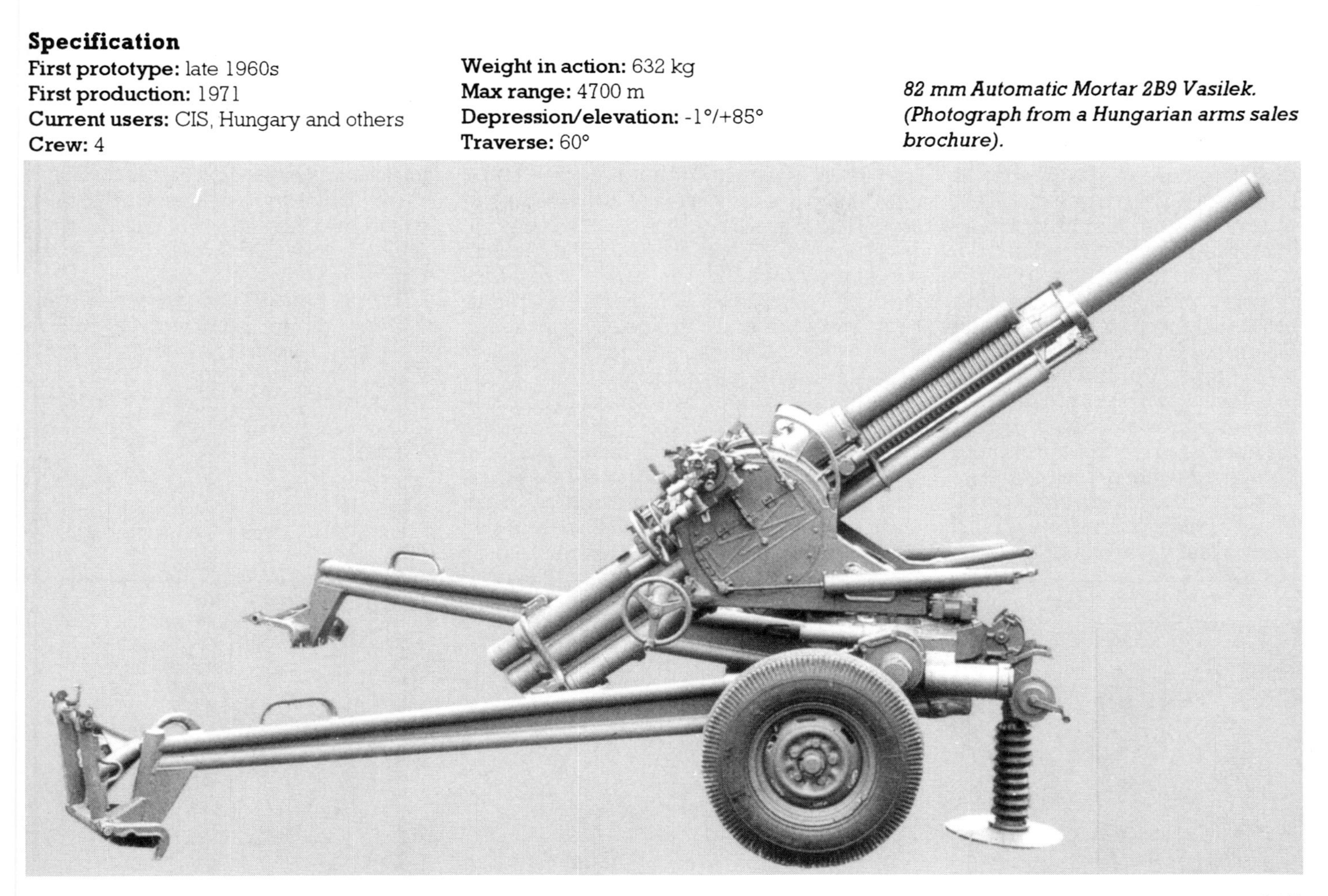

120 mm Mortar 2B11

The **120 mm Mortar 2B11** is an updated version of an earlier model, the 120 mm Mortar M-43, first produced in 1943 and destined thereafter to become one of the most widely used of all 120 mm mortars and a design which greatly influenced many others that followed - many M-43 mortars and their clones remain in production and in service with nations all around the world to this day (the Chinese 120 mm Type 55 is a direct M-43 copy). The **120 mm Mortar 2B11** is a revised version of the M-43, with most of the revisions involving modern materials which reduce the overall weight compared to the earlier model.

120 mm mortars are issued in place of other artillery support weapons with many infantry and other light formations in the Eastern Bloc armed forces. For transport the **2B11** is normally towed as a single unit on a light carriage, one of the main changes from the earlier M-43. Once at a firing site the barrel is simply lifted manually until the circular baseplate is on the ground. The carriage can then be removed for the barrel to rest on a bipod for firing. If required the barrel, baseplate and carriage assemblies can be separated for loading onto vehicles or for animal pack transport. In action, mortar bombs are manually muzzle loaded for the fixed firing pin at the bottom of the smooth-bored barrel to detonate the variable propellant charges - an alternative trigger mechanism can be employed. Some **2B11** mortars feature a muzzle device which prevents double-loading.

Several self-propelled carriages for the **2B11** exist, including a converted MT-LB tracked carrier used by Iraq and a Czech system involving a more formal locally-produced turretless variant of the BMP-2 tracked APC and known as the PRAM-S.

The **120 mm 2B11** can fire virtually any 120 mm mortar bomb (actual calibre 119.4 mm) produced in both the East and West. Maximum range is 7100 metres with minimum range being 480 metres. The rate of fire can be up to 15 rounds a minute.

120 mm 2B11

Specification (2B11)
First prototype: mid-1950s(?)
First production: 1960(?)
Current users: CIS, Hungary, Czech and Slovak Republics, Iraq, India, Poland and others
Crew: up to 6
Weight in action: 210 kg
Width travelling: 1 m
Max range: 7100 m
Depression/elevation: +45°/+80°
Traverse: up to 26°

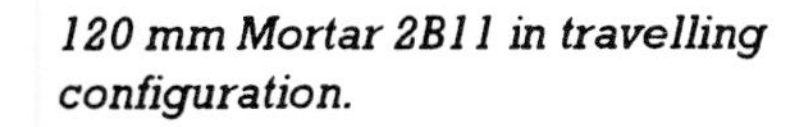

120 mm Mortar 2B11 in travelling configuration.

160 mm Mortar M-160

The 160 mm mortar originally entered service with the old Red Army in 1943, hence its original designation of M-1943. Originally introduced as a manufacturing expedient to produce artillery support weapons which required the minimum of scarce production facilities the M-1943 proved to be a great success. Later innovations, introduced post-1945, included a longer barrel, creating the **160 mm M-160** which was at one time widely used by all manner of Soviet Army and other Warsaw Pact units as a general fire support weapon. The **M-160** has now been withdrawn from general use but is still retained at division level by the successors to the Soviet Army mountain divisions where its relatively light weight compared with similar-calibred artillery pieces can be put to good advantage.

The **160 mm Mortar M-160** has a long smooth-bored barrel, so long that a breech loading mechanism has to be employed. For loading the barrel is pivoted about a central point so that once in the horizontal position a bomb can be inserted into the breech from a loading tray lifted by two of the seven-man crew. The barrel is then returned to the firing elevation angle. For transport the **M-160** is towed by a light truck or tracked carrier, using a muzzle attachment carrying the towing eye. The wheeled carriage remains attached to the mortar while it is in action.

Bombs fired by the M-160 are of only one type, HE. Standard HE bombs, of which two similar models are available, weigh 41.2 kg and can be fired to a range of 8050 metres. There is also a later model with a more streamlined outline and a steel body (rather than cast iron) which can reach 8500 metres but it is slightly lighter at 40 kg. The rate of fire is from two to three rounds a minute. There are seven propellant charges to choose from, a fixed primary charge plus six increments.

160 mm M-160

Specification (M-160)

First prototype: early 1950s(?)
First production: 1955(?)
Current users: CIS, Egypt, Iraq, India (offered for sale)
Crew: 7
Weight in action: 1300 kg
Length travelling: 4.86 m
Width travelling: 2.03 m
Height travelling: 1.69 m
Max range: (standard HE) 8050 m
Depression/elevation: +50°/+80°
Traverse: 24°

160 mm Mortar M-1943 drawn from a rare photograph.

240 mm Mortar M-240

The **240 mm Mortar M-240** entered service with the old Soviet Army during the early 1950s. One of the main reasons for its initial development was that it required considerably less production resources to manufacture than a conventional artillery piece of similar calibre. Limitations such as a general unhandiness and slow in and out of action times were accorded low priority since the **M-240** was usually issued to batteries concerned with the deliberate demolition of strongpoints or urban areas where mobility is relatively unimportant. The **M-240** is no longer a front-line weapon with CIS states but is still held in reserve.

The **M-240** uses a smoothbored barrel with breech loading. In action the barrel rests on a large and heavy circular baseplate on which is mounted a rectangular breech frame. The centre of the barrel pivots in this frame so that the breech block can be raised to the horizontal for loading. Once a bomb has been loaded the barrel is returned to the selected angle of elevation. The breech frame also carries the recoil mechanism.

Getting the **M-240** into action takes the eleven-man crew at least 25 minutes and almost as long again to withdraw - both operations require the use of hand winches on the carriage. On the move the **M-240** is towed muzzle first using a towing eye on a muzzle cover. Tractor vehicles are usually tracked carriers with other vehicles carrying the ammunition as each **M-240** bomb is 1.565 metres long and weighs 130.84 kg. The HE warhead contains 31.93 kg of TNT.

Maximum range is 9700 metres and minimum range 800 metres. The rate of fire is limited to about one round a minute, even with five men involved in the loading process and bringing bombs to the mortar on a two-wheeled trolley. At the breech the bomb is lifted by two men using a tongs lifting device while another man actually rams the bomb into the barrel and closes the breech block.

240 mm M-240

Specification

First prototype: 1949
First production: 1952
Current users: CIS, Egypt, Syria, Iraq, Romania
Crew: 11
Weight in action: 4150 kg
Length of barrel:5.34 m
Length travelling: 6.51 m
Width travelling: 2.49 m
Height travelling: 2.21 m
Max range: 9700 m
Depression/elevation: +45°/+65°
Traverse: 18°

240 mm Mortar M-240 being loaded.

240 mm Self-propelled Mortar SM-240

The old Soviet Army had great faith in the use of heavy mortars so when the **240 mm Self-propelled Mortar SM-240** was first observed in 1975 it came as no great surprise. Even so, the **SM-240** is rarely seen publicly, even though some 400 or so have been manufactured. The **SM-240** is also known by the name of Tyul'pan (Tulip Tree) or the designation 2S4 - in the West the **SM-240** was long known as the M-1975.

The **SM-240** is mounted on a tracked chassis with some similarities to that used by the 152 mm self-propelled gun 2S5 (qv). The mortar involved is known as the 2B8 and is mounted at the rear of the chassis in such a manner that the mortar and its large baseplate are hydraulically lowered to the ground for firing. The chassis also carries two crew compartments, one for the driver and commander and another for the other seven members of the nine-man crew.

In the firing position the smoothbored barrel points to the rear. Inside the hull are two drum-type magazines, each holding 20 bombs. Each drum can present a bomb to an open hatch which is aligned with the mortar breech after a mechanism moves the barrel to the loading position. Once the bomb is rammed home the barrel is then returned to the firing position. With this system the rate of fire can be one round every 62 to 77 seconds. Barrel movements are made hydraulically.

The **SM-240** fires its own family of 240 mm ammunition. The main bomb fired is HE-fragmentation weighing 130 kg. This bomb has a maximum range of 9650 metres but there is a further rocket-assisted HE-fragmentation bomb weighing 228 kg with a maximum range of 18000 metres, although accuracy suffers at the longer ranges. Other 240 mm bombs include concrete-piercing and a reported cargo warhead carrying PFM-1 scatterable anti-personnel mines. Tactical nuclear and chemical warheads are now withdrawn.

240 mm SM-240

Specification
First prototype: 1972(?)
First production: 1974
Current users: CIS, Czech Republic, Iraq, Lebanon
Crew: 9
Weight in action: 27500 kg
Length overall: 8.5 m
Width: 3.2 m
Height travelling: 3.2 m
Road range: 500 km
Fording: 1 m
Powerpack: V-59 diesel developing 520 hp
Depression/elevation: +50°/+80°
Traverse: 20°

240 mm Self-propelled Mortar SM-240 in travelling configuration.

120 mm MO-120-RT 61 Rifled Mortar

France

The **120 mm MO-120-RT-61** produced by Thomson-Brandt resembles a gun rather than a conventional mortar. For instance, the **MO-120-RT-61** has a rifled barrel rather than the usual smooth bore and is fired from its carriage wheels rather than from the usual bipod. However the **120 mm MO-120-RT-61** is muzzle loaded (although a trigger device can be involved) and is fired from a heavy baseplate at high angles of elevation.

Many armies use the **120 mm MO-120-RT-61** as a form of light artillery in place of conventional guns. Although the maximum range using standard HE ammunition is limited to 8135 metres this can be improved to 13000 metres when a rocket-assisted projectile (RAP) becomes involved, further than a standard 105 mm howitzer can achieve. When it is considered that the HE projectile involved weighs 15.7 kg on target and up to 20 rounds can be fired in one minute for short periods it becomes apparent that the **MO-120-RT-61** has considerable firepower potential.

Much favoured by airborne forces, the **120 mm MO-120-RT-61** can be towed behind a light 4 x 4 vehicle using a muzzle attachment with a towing eye. If required the weapon can be broken down into three main assemblies for other forms of transport. The 2.08 metre- long barrel, which is finned externally to dissipate heat during prolonged periods of firing, rests on its wheeled carriage during firing, using a cradle resting on a tubular bar between the carriage wheels.

Ammunition fired by the **120 mm MO-120-RT-61** has pre-rifled drive bands, with the projectile body more closely resembling an artillery projectile rather than the usual tear-drop mortar bomb outline. Propellant increments are added and subtracted around a post placed below the projectile base. Apart from the standard HE and HE-RAP already mentioned there is an anti-armour round with coiled steel bars wrapped around the warhead explosive to form the projectile body. On detonation the bars shatter into armour-penetrating fragments. There is also an illuminating projectile.

The potential accuracy of the **MO-120-RT-61** is such that it is often used in conjunction with a ballistic computer capable of delivering first-round-on-target results.

A prototype of a truck-based self-propelled version of this mortar, known as the MO.120-RA has been produced by LOHR SA.

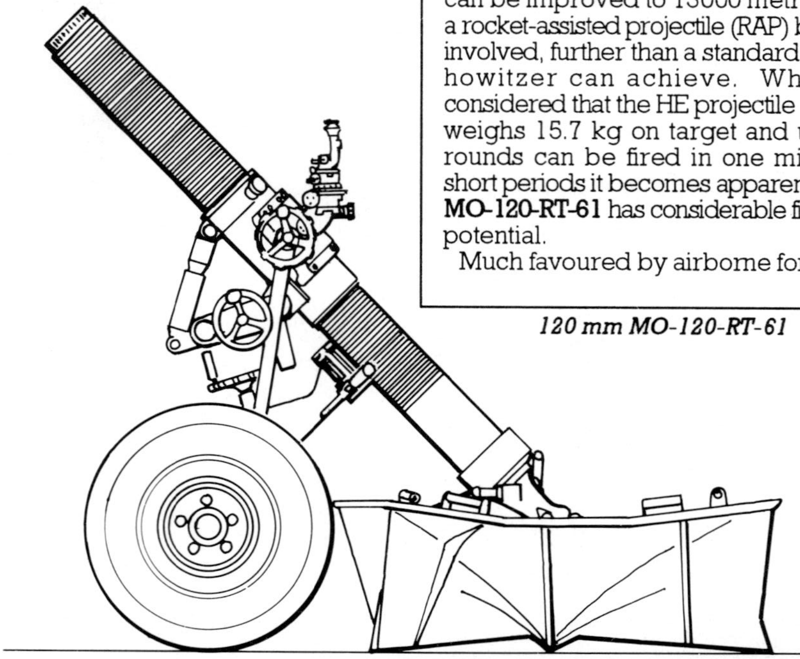

120 mm MO-120-RT-61

Specification
First prototype: 1959
First production: 1961
Current users: France and several other nations
Crew: 4 to 6
Weight in action: 582 kg
Length travelling: 3.01 m
Length of barrel: 2.08 m
Width: 1.93 m
Height travelling: 1.33 m
Max range: HE 8135 m; RAP 13000 m
Depression/elevation: +30°/+85°
Traverse: 14°

120 mm MO-120-RT-61 Rifled Mortar ranging.

120 mm Armoured Mortar System

UK

The **120 mm Armoured Mortar System (AMS)** has been developed by Royal Ordnance as a private venture from late 1985 onwards. Following early firing and other trials with a prototype mortar in a welded steel turret mounted on a M113 APC, a complete turret system was developed for mounting on an American Light Armoured Vehicle (LAV) with a view to its forming part of a large anticipated order from Saudi Arabia.

The **120 mm AMS** includes not just the mortar and its armoured turret on the LAV chassis, but a fully integrated fire control system as well. The combination provides a highly mobile fire support system on a relatively light vehicle.

The 120 mm mortar is breech-loaded from within the vehicle. Inside the turret the vehicle commander is seated on the right and the loader on the left. The smoothbored barrel with its fume extractor is 3 metres long and mounted on recoil buffers in a fully-traversing welded steel turret. Possible elevation angles are such that it is possible to use the mortar for direct fire or for firing at low elevation angles to prevent detection by enemy mortar-detection radars. A machine gun can be mounted on the turret roof.

Ammunition fired can be of virtually any 120 mm type from Eastern or Western sources. With most 120 mm ammunition the maximum range is about 8500 metres, rising to 12000 metres with rocket-assisted bombs. The normal rate of fire for sustained periods is four rounds a minute, with eight rounds a minute for a maximum period of three minutes. If required, bursts of three rounds in 15 seconds can be fired. The system can accommodate the 'next generation' of smart mortar rounds such as the Swedish Strix with its anti-armour, self-homing, infra-red seeking head.

Development of the **120 mm AMS** is virtually complete and although currently demonstrated on an LAV chassis, the complete system can be adapted for virtually any similar vehicle.

120 mm AMS

Specification (turret system only)
First prototype: 1986
First production: not yet in production
Current user: not yet in production
Crew: 2
Weight in action: approx 3000 kg
Length overall: 4.66 m
Length of turret: 2.59 mm
Length of barrel: 3 m
Width: 2.2 m
Height of turret: 0.89 m
Depression/elevation: -5°/+80°
Traverse: 360°

120 mm Armoured Mortar System with turret mounted on an LAV 8x8 APC.

Glossary

AFAS	Advanced Field Artillery System
AMS	Armoured Mortar System
APC	Armoured personnel carrier
APU	Auxiliary power unit
BB	Base bleed
BT	Boat tail
CIS	Commonwealth of Independent States
ERFB	Extended range full bore
ERFB-BB	Extended range full bore base bleed
FCS	Fire control system
FH	Field howitzer
FRAG-HE	Fragmentation high explosive
FS	Fin stabilised
HB	Hollow base
HE	High explosive
HE-BB	High explosive - base bleed
HEAT	High explosive anti-tank
HEP	High explosive plastic
HERA	High explosive rocket assisted
HESH	High explosive squash head
ICM	Improved conventional munition
LARS	Light Artillery Rocket System
LFH	Light Field Howitzer
LP	Liquid propellant
LTH	Light Towed Howitzer
MBT	Main battle tank
MG	Machine gun
MLRS	Multiple Launch Rocket System
NATO	North Atlantic Treaty Organisation
RAP	Rocket assisted projectile(s)
RD-MRWS	Rapid Deployment Multiple Rocket Weapon System
RO	Royal Ordnance
SRC	Space Research Corporation
T	Tracer
TACMS	Tactical Missile System
UFH	Ultralightweight Field Howitzer
VSEL	Vickers Shipbuilding and Engineering Limited